T0073805

MATH-*ISH*

Also by Jo Boaler

Limitless Mind: Learn, Lead, and Live Without Barriers

Mathematical Mindsets: Unleashing Students' Potential Through Creative Mathematics, Inspiring Messages, and Innovative Teaching

The Mindset Mathematics Curriculum Series: Visualizing and Investigating Big Ideas (K—8 books) with Jennifer Munson and Cathy Williams

The Mindset Mathematics Curriculum Series: Visualizing and Investigating Big Ideas (K 8 books) with Jennifer Munson and Cathy Williams

What's Math Got to Do with It?: How Teachers and Parents Can Transform Mathematics Learning and Inspire Success

The Elephant in the Classroom: Helping Children Learn and Love Maths

MATH-*ISH*

Finding Creativity, Diversity,
and Meaning in Mathematics

Jo Boaler

HARPERONE

An Imprint of HarperCollins*Publishers*

MATH-*ISH*. Copyright © 2024 by Jo Boaler. All rights reserved. Printed in the United States of America. No part of this book may be used or reproduced in any manner whatsoever without written permission except in the case of brief quotations embodied in critical articles and reviews. For information, address HarperCollins Publishers, 195 Broadway, New York, NY 10007.

HarperCollins books may be purchased for educational, business, or sales promotional use. For information, please email the Special Markets Department at SPsales@harpercollins.com.

FIRST EDITION

Designed by Bonni Leon-Berman
Illustrations by Kane Lynch

Library of Congress Cataloging-in-Publication Data has been applied for.

ISBN 978-0-06-334080-0

24 25 26 27 28 LBC 6 5 4 3 2

*This book is dedicated to my niece Imogen (1995–2021),
and to Julie, Vic, and Alex. Imi, you are always with us.*

CONTENTS

1

A NEW MATHEMATICAL RELATIONSHIP

I had been invited to a dinner in an expensive restaurant to meet the CEO of a major social media platform and his wife. The restaurant was lavish, typical of the expensive venues in Silicon Valley. I felt nervous as I joined them at the table, wondering what the evening had in store. The occasion had come about through a friend who knew the CEO's wife. My friend was familiar with my work to improve mathematics teaching and thought the CEO would be a useful person to connect with. After living and working in Silicon Valley for the past several years, I understood that this kind of networking is part of the fabric of the area and a large reason for the growth in innovation and productivity here.

The beginning of the dinner was baffling, unlike any I had ever attended: the CEO acted as though I, and the rest of the group he was eating with, were not there. He instead spent all the time on his phone, talking to colleagues, busily making work plans with a stack of work papers he had pulled out of his bag. This behavior, intentional or not, made us all look and feel insignificant. His wife appeared embarrassed, glancing repeatedly toward her husband's makeshift office at the corner of our table.

This continued until the food arrived and the CEO was forced to pack up. We were halfway through the meal before he acknowledged my existence. Looking up from his food, the CEO stared intently at me and said disapprovingly, "So you think math teaching should change?"

Without pause, he went on to tell me how well he had done in math, revealing his many achievements in mathematics in school and in college. At this point, I knew the conversation would not be easy. After many years of trying to improve mathematics teaching—a subject with widespread failure rates—I knew that those who had achieved highly are usually the ones believing nothing should change. In their minds, math is hard, and they had just showed their brilliance by achieving at high levels. But one thing you will learn about me is that I am willing to fight for solutions to what I know are real problems for many students. I decided to show the CEO a different mathematics.

I explained how neuroscientists have given us insights into how our brains process mathematics—and how important it is

1.1 The author showing Ruth Parker's increasing squares model, shown in figure 1.2

that we activate different parts of the brain in our mathematical thinking, particularly the visual pathways. He agreed to look at an interesting visual with me, which I often share when meeting someone new. I chose one of my favorites, created by the mathematics educator Ruth Parker (figs. 1.1 and 1.2).

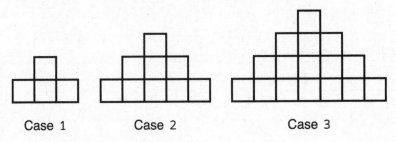

Case 1   Case 2   Case 3

1.2 Ruth Parker's increasing squares model

Typically pictures like this are used to help students think about pattern growth and, from there, generalization using algebraic symbols. Students in math classes are often asked questions like "How many squares would be in case 10? Or case 100? Or case n?" These are good questions, but they become much better when they invite visual thinking. Typically, in classrooms, students are expected to draw a table of numbers, then stare at the table until they notice the pattern. The pattern that people may notice is numerical—that is, to find the number of squares in the pattern, you can take one of the case numbers (such as 2), add one to the number (making 3), then square the number to get 9. This pattern of adding one and squaring the number lets you find the total number of squares from any of the case numbers. This pattern can be expressed algebraically as $(n+1)^2$.

The expression that describes the pattern, $(n+1)^2$, is a quadratic function. When students work in this way—manipulating numbers and symbols with no connections or meaning—they miss important opportunities to understand mathematical functions. In my work, instead of asking how many numbers are in different cases, I ask, "How do you see the pattern growth? Where do you see the extra squares on the shape?" These are the questions I asked the CEO that night.

Number of Squares in Each Case

Case	Total number of squares
1	4
2	9
3	16
4	25
n	$(n+1)^2$

I was surprised by his response. It was not that the CEO could not see a method of growth; he could, and he described the method to me. He said that he saw the extra squares on top of each column. Other people have described this as a "raindrop" method, with the squares coming down on top of the shape, like drops of rain coming down from the sky. Figure 1.3 shows this method and other ways people see the growth.

But after sharing his method, the CEO asked something I have never been asked before. With genuine confusion in his voice, he asked, "Doesn't everyone see it that way?" I did not tell him no; I just asked everyone at the table to share how they saw the growth. As we went around the table, everyone shared a different way of seeing the pattern growth. The CEO appeared increasingly flabbergasted, as though it had never occurred to him that there was more than one way of seeing something mathematical. He shook his head in disbelief. We had his attention.

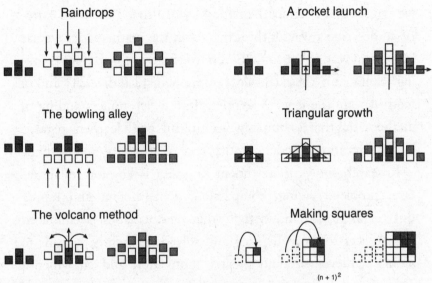

1.3 Different ways people see and describe the pattern growth

Changes to questions to invite a broader version of mathematics are important. When students encounter the narrow, numerical version of a question, and they stare at tables of numbers to see patterns and find algebraic expressions, they may come up with $(n+1)^2$, but they have no idea why this expression works or what it means. When we ask students how they see the pattern growth, they understand the function more deeply, and they can understand, visually, that the shape grows as a square that is always one more than the case number. The last method in figure 1.3 shows this most clearly. This is why we can describe the growth as $(n+1)^2$.

As dinner progressed, I shared something I am passionate about, which is rooted in important neuroscientific research: the value of mathematical diversity. The term *diversity* means difference, variety. In this book, I will use the term *mathematical diversity* to include both the value of diversity in people, whether racial, cultural, social, or any other form, and the diverse ways we can see and learn mathematics. I will also use the term *math-ish* to describe a way of thinking about the mathematics we use in the real world that could be a powerful tool for student thinking. Embracing these concepts of mathematical diversity and of math-ish are the keys to stepping into a rich understanding of mathematics that is equitably meaningful for all learners, regardless of education, gender identity, race or ethnicity, and beyond.

Research shows that student diversity is key for collaboration, problem-solving, compassion, achievement, and more.[1] But research also shows that when mathematics is embraced as a subject that can be seen and solved differently, it leads to higher achievement and greater motivation and enjoyment.[2] These two aspects of diversity—differences among people and in mathematics—stand alone, but they also come together in beautiful ways, strengthening and supporting each other. If we want to value and encourage all the people we meet who think in different ways, we must reject narrow mathematics, which is the only mathematics most people know. Instead, we should embrace mathematical diversity.

The CEO was amazed that night by the mathematical diversity he saw, which is often missing in schools and homes, much to the detriment of individuals' mathematical relationships. Some people can be successful with a narrow and one-dimensional version of mathematics, but even these people are missing out on the full range and power of math. When people engage with the diversity of mathematics, it changes the insights they gain from every numerical, spatial, or data-related situation they encounter.

A DIFFERENT WAY

I am a professor at Stanford University, but I started my career as a mathematics teacher in London schools. I first taught at Haverstock School, a secondary school in Camden Town in central London.[3] Camden is a vibrant, beautiful, and under-resourced part of London, and most of the students live in assisted housing and qualify for free school meals. When I was teaching at Haverstock, the students spoke more than forty different languages. It is a wonderfully diverse place to teach.

On my first day of teaching there, I was fresh from my teaching credential year at London University and filled with ideas on ways to bring mathematical beauty and joy to students. The students in the class were thirteen years old and had just been put into ability groups. I was teaching the lowest group, "set 4" of four groups. Here I met Sue, a feisty student who I later learned had a reputation for being excluded from school. Her vocal opposition to some of the teachers' ideas meant she was forced to miss school days—and learning opportunities. Sue wore her trademark cheeky expression and a twinkle in her eye that first day when she asked me, loudly, "Why should we bother?"

I hesitated. As a brand-new teacher in the first hour of my first class, I was unsure of what to say. Her pointed question was valid. Students placed into low groups in the UK system can only achieve low exam grades. The highest grade these students in my class could achieve in their national examinations,

1.4 My first day of teaching, Haverstock School, London

which would take place three years later, was a D. Most jobs and higher education pathways require a C or above. If students hear a strange, loud noise at the time they are placed into low groups, it is probably the sound of doors slamming—doors that could lead to brighter futures but which are no longer available to them. In that moment, I decided I would teach Sue, and her peers, the higher-level work. Three years later, Sue achieved the grade she needed to advance and gained a place in a sound engineering program. She now owns and runs major music and entertainment companies in Bali.

When Sue first came to my class, she held the idea she could not do well in mathematics, and she was facing many difficult circumstances at home and at school, including being tracked into the lowest group. Despite this, she was able to transform her mathematical achievement—and, with that, her life. In a news story on her achievements in later years, she reflected that prior to her achieving well in mathematics, she had been thinking she would never accomplish such success in her life.

In the years since then, I have taught many people what I taught those students at Haverstock School in London—a way of approaching mathematics that leads to success. This starts with mathematical diversity—valuing the different ways that people see and think about mathematics. This alone has the potential to change mathematics from being a narrow, rigid experience to a diverse, accessible, and dynamic experience. It also involves taking an "ish" approach to mathematics, but I will explain more about that later.

NARROW MATHEMATICS

Many people are aware of the damages caused by the lack of mathematical diversity in the school system, what I call narrow mathematics. In the world of narrow mathematics, questions have one valued method, and one answer. They are always numerical, and they do not involve visuals, objects, movements, or creativity. Most people have only ever experienced narrow mathematics, which is why we have a country of widespread mathematics failure and anxiety.[4] One example of the damage caused by narrow mathematics comes from the college system. In a noteworthy *New York Times* article, investigative reporter Christopher Drew shared that every year, students enter four-year colleges intending to major in one of the much-needed STEM subjects of math, science, engineering, and pre-medicine.[5] After the students experience the introductory classes, however—what Drew describes as a "blizzard of calculus, physics and chemistry," a stunning 60 percent of the students change their majors. Drew quotes David E. Goldberg, an emeritus engineering professor, who describes this as "the math-science death march."

Drew cites as an example Matthew, who scored an 800 in his math SAT and had taken calculus BC and five other advanced placement courses in high school. He arrived in college intending to become an engineer, expecting to encounter interesting content that he could think about in different ways. Instead, he found himself in classes where he was expected to memorize equations and decided, at that point, that he had "had enough" of narrow math. As a potential engineering student, Matt had expected the classes to be "all about the application." He was so disappointed with the narrow ways in which content was introduced, he switched his major to psychology, where the courses invited students to discuss ideas.

Narrow mathematics not only pushes high-achieving students away from STEM courses but also has a devastating effect on students who need to pass mathematics to move forward with their lives, whatever direction they intend to take. Approximately 40 percent of students in the US attend community col-

1.5 The math-science "death march"

leges, where they are required to take tests, usually in the content of algebra 2. Eighty percent of these students end up in remedial math courses, typically consisting of algebra taught in the same way they experienced it in high school, which caused their failure in the first place. In California, more than 170,000 students are placed into remedial math, and over 110,000 of them fail or leave the courses and cannot move forward with their college careers.[6]

Narrow mathematics ends the hopes and dreams of millions of college students. Not only is this a problem for the students, but it also causes severe problems for US society, threatening the future of the economy, the development of science, technology, medicine, and the arts.[7] Indeed, the data is so dramatic, negative, and consequential, I am surprised it does not prompt federal- and state-level action to illegalize narrow mathematics, banishing it from K–12 and college classrooms.

The problem highlighted by the college data from two- and four-year colleges is reproduced in classrooms throughout the country, from kindergarten all the way to college: very few students enjoy or connect with a version of mathematics that is a

narrow, impoverished version of the subject. As students move through the grades, mathematics becomes more and more narrow, and the narrowness of the subject is mirrored by the narrowness of the students who manage to stay engaged and successful.[8]

When we open mathematics to acknowledge the many ways that any mathematical thought or concept can be considered, seen, and solved—when we teach mathematics with diversity—we open the subject to many more students.[9]

Even from the very beginning of my career, I knew there was a better way to teach and learn math. But when the new neuroscience of learning exploded onto the scene ten or so years ago, showing the ways our brains process math, I became passionate about communicating these ideas.[10] When I share ideas with others, I do not present abstract neuroscientific results; I translate the findings, showing what they mean for mathematics learning—and, more broadly, mathematical relationships. The ideas have the potential to transform learning and so are incredibly helpful for parents, teachers, and students. But in addition, they also change the way people encounter and use mathematics

in their lives. Mathematics can be a secret weapon, an incredible tool, that all of us have the potential to use, but it is often an underutilized skill. If you would like to live your life to the full, making the most of the mathematical lens we can use to see the world, I invite you to power up, approaching mathematics and life with the lens of mathematical diversity and math-*ish*.

A GLOBAL CULTURAL PROBLEM

The ideas I will share in this book have helped not only those who did not achieve well in math. I have been teaching highly accomplished undergraduates at Stanford for years now. Most of them arrive with a broken relationship with math. They have been successful, but they see math as a set of procedures they need to reproduce, at speed. When I show them that math can be the opposite of this—a set of connected and creative ideas that people can think about slowly—they are amazed, and thrilled.[11] The students tell me that they do not ever want to go back to the narrow, speed-based version of mathematics they knew before.

Few people see and experience mathematics with diversity, and the consequences of bad math are real for millions of people worldwide. Between 10 percent and 40 percent of people in most countries are innumerate and avoid math as much as possible.[12] These folks are vulnerable every time they need to read a chart, mathematical diagram, table, or set of numbers. Many in this position are experiencing poverty, and the inequities in education systems and society deny them opportunities to learn and improve their lives. Unfortunately, the people who most need mathematical confidence and knowledge are often those who are not given access to a strong mathematics education, so they are cut off from many career paths.[13] Mathematics achievement can help bring young people out of poverty and allow them to live their lives more fully.[14]

In addition to the lack of diversity in classrooms, many people have a negative relationship with math because, more than any other subject, it is treated as a performance subject in school. Math is the most over-tested subject in the curriculum, and it is frequently used to rank students and, by extension, to measure their worth as people. Students often do not even think about math for its own sake; they can only think about how well they are doing in the subject. To make it worse, testing usually consists of cold, narrow questions taken at speed.

Very few people survive these narrow, speed-based testing assaults with positive ideas about the subject. Math testing is a trial by procedures. For those who do well, their reward is to meet real mathematics—to play with ideas and make connec-

tions between them. But for most people, their experience of procedures, pressure, and constant judgment leads them to conclude that math is scary and unpleasant.

If you are a parent or teacher with students facing these issues, I have good news: you are not alone, and this book will help you. The vital information in these pages will help children navigate their mathematical journeys, finding joy, relief, and hope in this diverse approach to mathematics. Even if you are an adult with a positive relationship with math, this book will help you, too, as it will discuss new research about the mathematical approach we need in the modern world. I have worked with many people over the years who were able to change their mathematical relationship—realizing that it was not they who were "broken" but the system. When they changed the way they related to math, they found their lives enhanced.[15]

One of the mathematical powers I will teach you is to change any difficult mathematical problem into an easier problem. Most people do not know how, or think they are not "allowed," to do this. They think they must work with the difficult problem they are given. But changing math problems turns out to be really helpful, not just in school but in the demands that come our way in the world. This different way of thinking about mathematics is a new lens that can be applied to anything—and once you learn this superpower, you will not look back. I will also teach you to take the knowledge you have and make connections to other knowledge. If this sounds mysterious, read on, because it will all become clear in the chapters that follow.

The mathematics performance culture of schools became evident to me a few years ago, when I interviewed passersby in the streets of San Francisco, California. I was preparing to make one of my first online courses, and I arrived in the city on a cool day with a couple of my graduate students and some cameras.[16]

I asked each person who passed by to simply describe math to me. But they all answered a different question. One by one, they all shared how well they had done in math—they described their performance, usually ranking themselves as they talked.

Their emotional descriptions of their performance journeys, whether successful or not, were striking. Not one person gave a description of mathematics; instead, they described their own journeys of success or failure. This is the damage of the performance culture; it robs people of their right to enjoy mathematical ideas and become mathematically empowered.[17] Instead, many people have only ever known math as a tool for ranking, judging, and segregating.

It is not only the over-testing of math that creates this damage; the testing culture combines with the traditional misrepresentation of the subject as a set of procedures and right and wrong

answers. It is not a coincidence that those aspects of math appear together. It takes work and effort and some imagination to create test questions that assess a diverse approach to mathematics—valuing students' thinking, creativity, and problem-solving. None of the major math textbook publishers, the math app developers, or the testing companies have ever made this commitment. They love narrow, procedural math because it is easy to reproduce mindless questions in hundreds of thousands of textbook pages, and it is easy to test. But this should not be what drives mathematics education.

On the stage of misfortune, we have, so far, two villainous characters: an over-tested subject and a misrepresentation of math as a set of procedures. These two villains are bad enough, but they are helped by a third: the nasty ghoul of the "math brain."

For centuries, many believed that some people were born with a "math brain" and could learn math to high levels, while others could not. Often these ideas were combined with sexist, racist, and other discriminatory ideas about who had these special "gifts."[18] But the past ten years or so have shown, quite defini-

tively, that there is no such thing as a math brain, and all brains are constantly developing, connecting, and changing.[19] This is supported by neuroscientific evidence revealing incredible brain growth after short interventions.[20] It has also been demonstrated by people whose early years of school were difficult, often being labeled as needing severe special education support, but who went on to achieve the highest mathematical levels—including a doctorate in applied mathematics from Oxford University.[21]

These three villains—the math brain, the procedural subject, and the over-testing—with some systemic inequalities thrown in for good measure, all work together in perfect synchrony to create a horrible experience that few recover from. To make matters worse, some of the people who have come through this harrowing experience, usually people who are wealthy and powerful, fight with those trying to change it, to keep it the same. They survived it; why not make others try to do the same? I refuse to give in to these people, and my continued resistance has resulted in more than a few war wounds over the years.[22] My most recent experience of being one of the writers of a new mathematics framework for the state of California led to my being targeted with hate mail and death threats from a group working to discredit my research and work.[23] I continue the struggle because I know that mathematics can be experienced in a wholly different and beautiful way—a diverse way. And when people experience mathematics in this new way—even inside the brutal performance culture that pervades schools—math enjoyment and high achievement result, even on narrow tests.[24]

Fortunately, not all the people who have been successful in our math performance culture fight to keep it the same. I am heartened by the collaborations I have had with some incredible mathematicians, engineers, and scientists who know that we need to fundamentally change the ways mathematics is taught and learned.[25] Eugenia Cheng is one mathematician I admire who is devoting much of her career to writing to try to change the public's ideas of what mathematics is and can be. She accurately points out that we do not spend enough time sharing mathematical joy with young people; instead we focus on training them to jump through arcane hoops, encouraging them to memorize methods and rules that will probably have little or no use in their lives.[26]

THE MINDSET-COGNITION LINK

There are many reasons why it is important to *think about* math differently. Lang Chen, a neuroscientist and professor at Santa Clara University, conducted important research in this area.[27] It had been known for some time that students with positive attitudes toward their learning achieved at higher levels.[28] Positive attitudes reduce anxiety about learning, enhance motivation, and boost students' persistence.[29] Chen was interested in exploring this relationship further to determine the neurological mechanisms at play and the factors that support or impede a positive attitude.

One of Chen and his colleague's findings was that students' attitudes toward math—whether they liked or disliked it—correlated with their math achievement (but not their reading achievement), even after controlling for IQ scores,[30] age, working memory, and math anxiety.[31] Significantly, they found that positive attitudes were related to activity in the hippocampus—that is, the activation of the right and left hippocampus regions. This is important because many people—inside and outside the scientific world—believe that people's attitudes are unrelated to their mathematical cognition, that they exist in some other, fuzzy part of the brain. But the hippocampus is one of the most mathematical regions of the brain, playing a vital role in learning and in spatial navigation. The hippocampus has Google-like abilities; neuroscientist Sian Beilock describes it as the "search engine of the mind."[32] We can switch this power on for students if we spend more time encouraging them to enjoy mathematics through mathematical explorations that are absent of testing, grading, and other performance pressures. Chen found that what you believe about math changes your hippocampus—it changes the ways your brain functions when you learn.[33] So much funding and attention has been given to changing students' knowledge and understanding, but hardly anyone pays any attention to the important fact that huge boosts in achievement (and more) come about when students' attitudes and feelings change—their mathematical mindsets.

Scientists have found that when people with math anxiety are given math questions, the same fear center lights up in their brains as the fear center that lights up when we see snakes and spiders.[34] Considerable evidence exists showing that fear and anxiety disable parts of our brains, including the hippocampus, which reduces learning. Conversely, positive thinking and beliefs about mathematics boost the same important parts of our brains, leading to positive learning and higher achievement. This evidence alone should halt the use of anxiety-inducing practices in mathematics classrooms and encourage an approach that infuses mindset and positivity into the content. This is how we teach middle and high school students in our summer camps at Stanford. The impact is significant.[35]

The first summer camp we taught was for middle school students. Achievement tests showed that the four weeks they spent with us resulted in gains equivalent to 2.8 years of school.[36] The students changed their mindsets, and they started to believe in their own potential. They also began to see mathematics differently. The combination of these changes, in their mindset and their mathematics approach, was powerful.

The results were so strong that my youcubed team at Stanford—all experienced educators—began to run workshops for other teachers to offer the same camps in their local regions.[37] A study of the impact of these camps, enacted in ten school districts across the US, showed the same beneficial results in achievement.[38] Not only did the students achieve at significantly higher levels at the end of their camps, but they also went back to their schools and achieved significantly higher math GPAs the following term, compared to control students.[39] What happened in these camps that changed the students' learning trajectories?

They learned to approach mathematics in the ways I will share in this book.

A NEW MODEL FOR SUCCESS

My own knowledge about educational approaches that bring about transformational experiences for students comes from a wide array of sources. I work with and learn from neuroscientists at Stanford, and they have provided incredible and important insights in recent years, sharing how our brains process mathematics.[40] I also learn from cognitive psychologists who study learning and have shared important new knowledge about thinking and learning. Scientists such as Carol Dweck, Anders Ericsson, and Jim Stigler have written books sharing invaluable wisdom from psychology.[41] But I know from many years of teaching that while these neuroscientists and psychologists add critical insights into learning, the potential of their ideas is only maximized when it is combined with the insights that educators provide. For it is educators who have rich understandings of students' learning in classrooms and homes, and it is researchers in education who have studied varied learning environments to understand the impact of different teaching and learning approaches. Mathematics understanding is a basic form of literacy that every person needs—and that every person can achieve. In this book I do something I have never done before; I bring together the different insights these diverse sources provide into one model of teaching and learning. It is this model that I and others have used for the past few years that has resulted in significant and lasting achievement boosts.[42]

...

Popular books in mathematics education share classroom structures and ways of organizing classrooms, which can be generative for student thinking and learning and helpful for teachers, but they often leave out the building of important mathematical relationships.[43] These relationships—between adults and students, between students and students, and between you and you—matter. I know that positive mathematical relationships come from two important areas of work: one involves changing the mathematics people meet, from narrow questions to open questions that invite diverse, creative ideas; the other involves encouraging respectful, collaborative relationships between people. This book is filled with information on ways to bring about both conditions and inspiring examples of those who do so, with stunning results.

I hope that the ideas in this book help you, and those you work with, achieve beautiful mathematical relationships. Whether you are a parent, a student, an educator, or a person who would simply like to improve your relationship with math, I invite you to meet, or become reacquainted with, two concepts that will come alive in the following pages: mathematical diversity and math-ish. These ideas take place when the diversity provided by different people and ideas combines with a version of mathematics that is open and *ish* enough to benefit from that diversity. If you meet and truly experience this version of mathematics, it will change you. Then, perhaps, if you are ever stopped on a street and asked to describe math, you won't talk about your brutal experiences or your ranked performance; instead you will describe the ways math lights up your world.

2

LEARNING TO LEARN

It turns out that the most mathematically empowered people in the world take an approach to learning math that is different from those who are less successful. They do not achieve highly because they were born with special advantages but because they have been given access to some ideas and ways of working that I will share in this chapter. Usually, they have learned these ideas from friends or family members, as they are rarely shared inside the school system. But when learners are given information that allows them to approach math differently, their learning pathways change.

Importantly, the different, more successful approach to understanding math can be learned. And the impact of the changed approach is far-reaching; research shows that the practices and beliefs that create higher-achieving students are also practices and beliefs that improve people's lives. The changed approach starts with metacognitive actions, which allow people to become better problem solvers, communicators, and questioners; become more motivated; develop better relationships; and become more successful in their jobs. For learners, the impact is just as impressive, giving them huge boosts in achievement, a benefit that extends across all age groups and all subjects.

In my work with educators, I find that most people believe metacognition is a process of thinking about your own thinking. I prefer to think of it as learning to learn, and learning to be effective in life. Metacognition is a major part of developing into a creative, independent, self-regulating, and flexible problem solver. Despite the evidence supporting the impact of metacognition practices on learning and achievement, I rarely ever see them enacted in mathematics classes. I am doubtful that the ideas are regularly encouraged in workplaces or homes. This absence of metacognitive learning has major ramifications on long-term individual success in mathematics and beyond. Fortunately, there is a set of strategies that we can all use. As I share them in this chapter, I invite you into a metacognitive journey that will enhance your learning and your life.

A NEW THEORY OF COGNITION

In 1979, Stanford professor of psychology John Flavell created the theory of metacognition, and researchers have been investigating its impact ever since then.[1] The word *meta* comes from the Greek prefix meaning beyond, and metacognition regards the important processes that go beyond thinking, such as planning, tracking, and assessing. Flavell describes metacognition as including knowledge of ourselves, knowledge of the task at hand, and knowledge of strategies, so it is no surprise that it boosts problem-solving, enables mathematical diversity, and enhances work performance.[2] I am sharing the strategies that lead to a metacognitive and mathematically diverse perspective in this early chapter, so you can employ them as you move through the book.

In 2015 a wide-scale Programme for International Student Assessment (PISA) considered the learning approaches of 15 million students and the ways they related to mathematics achievement. The results showed that students taking a memorization approach to math were the lowest-achieving students in the world. The students who were the highest achievers were those taking a "relational" or a "self-monitoring" approach.[3] These students either described their approach to learning as focused on relating ideas to each other (see chapter 6) or described how they monitored their own learning; both approaches are quintessentially metacognitive. The Organisation for Economic Co-operation and Development (OECD), the international group that organizes and administers PISA tests worldwide, has set out a "learning compass 2030" defining the "knowledge, skills, attitudes, and values that learners need to fulfil their potential and contribute to the well-being of their communities and the planet."[4] Metacognitive skills are central in the guidance they give in developing students into people who function effectively in, and can contribute in meaningful ways to, their local communities and the world.

John Hattie is a researcher who conducts meta-analyses, an approach that combines multiple different scientific studies to find results that have wide-scale applicability. Hattie conducted a groundbreaking study that considered different approaches in education, each of which was given an effect size. An effect size is a value that communicates the strength of relationship between two variables—Hattie looked at the relationship between different approaches in education and student achievement, seeing how much the different approaches impacted achievement. He considered 138 different educational approaches through

70,000 studies and 300 million students and found that the average effect size (measured as a Cohen's d) of the different approaches was 0.40. He then decided to judge the effectiveness of them all in relation to this hinge point to see which approaches are worth pursuing. One of the approaches leaped out as having more impact than any other—with an effect size of 1.33: students reporting their own progress. The other acts that were shown to have high impact were classroom discussions (0.82), students engaging in metacognition (0.75), and problem-solving teaching (0.68). Among the approaches that had an impact size so low they did not justify their use were individualized instruction (0.23), external accountability systems (which I think of as district-imposed tests!) (0.31), and ability grouping (0.12).[5] Hattie's meta-analyses summarize thousands of studies, so they do not give detail on different ways the approaches are enacted, but they do give statistically powerful clues as to the approaches that are important to use.

Only one of the categories was called "metacognition," but the act that has the greatest impact of all—students reporting their own progress—is highly metacognitive; later in the chapter I will share some ways of giving this important opportunity to students.

In classrooms and workplaces, it is easy to spot people who have learned metacognitive strategies and others who have not. We probably all know people who are discouraged when they are given difficult challenges, assume that they cannot do well, and give up in the face of roadblocks. We probably also know people who are not open to ideas different from their own and who try to shut other ideas down, or perhaps they shut down themselves when different ideas are shared.

By contrast, a person who has learned metacognitive strategies is likely to be inquisitive and curious, they are eager to learn, and they appreciate diverse viewpoints. If they are stuck in a problem, they may circle back and think about what they know and need to know, or they may choose from other different strategies they have learned. Importantly, they enjoy the process of problem-solving and learning. This complex combination of high-level problem-solving, mindset, and planning that occurs when we are metacognitive takes place in the anterior prefrontal cortex of our brains.[6]

When people learn to *be metacognitive*, they not only improve their problem-solving, they also develop greater prosocial behavior, become better communicators, develop more empathy, and learn greater executive control.[7] Some people learn to engage in these different ways, invoking a complex combination of mindset and higher-order thinking through their lives, and they are more successful people because of it.[8] As Donna Wilson and Marcus Conyers, two experts on "brain-based learning" say, if cognitive functions are musicians, metacognition is the conductor.[9]

I see the potential of taking a metacognitive approach in three different areas of life and work. First, the area most often associated with metacognition is the self-awareness we have of our own learning and interacting. In my work teaching teachers in the UK and the US, I often ask them to reflect on a lesson they have taught. This has shown me something fascinating. Some teachers are incredibly reflective and recall details of lessons and ways they could have engaged differently, dissecting important classroom interactions and considering their role within them. Others don't reflect on much at all, only saying lessons were okay or went well. Unsurprisingly, it is always the reflective folks who become more effective teachers and often go on to become leaders in education. Importantly, this self-reflection is something that we can all learn and can be encouraged by teachers, parents, and others.

A second aspect of metacognition involves different ways of focusing on the task at hand, being willing and able to unpack it and think about what is involved. A metacognitive person will think in important ways—possibly going back to the question, considering what information is needed, thinking out loud, or simplifying the question. Someone who has developed and reflected on different strategies can choose among them, or try a few different approaches.

The third part of metacognition involves assessment, being able to track one's own progress and reflect on what is needed to achieve goals. This is where teachers and parents play a critical role in setting out for students where they should be going and ways to get there. Education leaders Paul Black and Dylan Wiliam proposed an approach, which they called "assessment for learning," defining it as communicating to students where they are now, where they need to be, and how to close the gap between the two.[10] This information, which creates responsible learners who regulate their own learning, is not typical of assessments used in mathematics classrooms.[11] Later on I will share cases of students engaging with this important opportunity.

PUT METACOGNITION INTO PRACTICE

When I work with groups of teachers, I often find that they are fully aware of the value of being metacognitive, but they are less aware of ways to develop this approach in themselves or in their students. In the next section I will share some of the most impactful approaches that I have used or seen used, and the responses from the people who have experienced them. It is not an understatement to say that for many students, learning the strategies I share below unlocks their potential from that moment on.

Share the Value of Metacognition

The logical place to start people on a metacognitive learning journey is by sharing the importance of different ways of interacting with knowledge that are available to us all. Many researchers have shown the impact of metacognition on students' achievement.[12] It is important to share with learners that there are different ways they can engage with mathematics, and other ideas, and that those different ways of engaging matter.

Donna Wilson and Marcus Conyers are experts in brain-based approaches to learning, and one strategy they share for younger students is to invite them to draw or decorate their own brain cars, telling students that they can be "driving their own brains."[13] Students can think about steering their car away from distractions or backing their car up to reconsider directions.

But communicating the importance of being metacognitive only helps if we also encourage the ways we can be metacognitive—the different strategies and ways of thinking and communicating that people can use. I recommend letting

people know that engaging metacognitively matters and then helping them learn a range of metacognitive strategies.

Over the years, I have been fortunate to learn from some amazing teachers of mathematics who have developed metacognitive learners. One of those teachers is Carlos Cabana. He uses teaching strategies to help foster metacognitive learners during whole-class discussions and at other times. We will begin our consideration of different metacognitive approaches that we can all share and use in our lives by delving into a teaching example from Carlos.

Encourage Metacognition Through Discussion

I have known and admired Carlos Cabana for many years, and I have been fortunate enough to study his teaching approach in multiple schools. When I studied the students at Railside School, a diverse urban high school where Carlos was the department cochair with Lisa Jilk, I witnessed high school students who were highly effective problem solvers because teachers taught them how to work well together.[14] When students sensed someone in their group was not contributing or working well, they would invite them into the discussions; when students did not know how to start on a problem, they would ask each other reflective questions—such as "what is the question asking us?" As students worked through the problems they were given, they would try different strategies that they agreed on together.[15] It is my belief that many teachers—or workplace managers—would watch the Railside classrooms and be wowed by the students' ability to work independently of the teachers and to work so well together. I have no doubt that the students' learned practices of working well together played a role in their high achievement, which was significantly higher than students learning at the other, more typical schools I was studying.[16]

While the students at Railside were learning to be effective problem solvers, they were also learning "relational equity," a form of equity that is not about equal test scores but respectful relationships between students.[17] One of the most important goals we should hold for our students learning to be citizens of our diverse world is this form of respect for their fellow students, no matter their race, class, gender, achievement level, or any other form of difference.[18]

During the first week of school when Carlos was instructing sixth-grade students, he demonstrated how educators can encourage thoughtful metacognition in their students. Watching a teacher in the first week of the school year is always incredibly informative, as this is the time when teachers establish what will be the norms of their classrooms for the year ahead.[19] When I observed Carlos in his first class teaching sixth grade, it became clear to me how his students became aware, effective, and metacognitive problem solvers. I noticed, for the first time, that every instruction he gave the students was an invitation to reflect metacognitively.

Inside Carlos's Classroom

Carlos began his lesson by asking for a volunteer to draw a rectangle with twelve squares on the board at the front of the room. As this was the first mathematics request of the year for the sixth-grade students, Carlos was particularly aware of the need for him to value the students' thinking, to show meaningful understanding, clarity, and tenderness. Ana volunteered and nervously approached the board. She was the first student in the class to present that year. As Ana went to the front, Carlos asked the other students, "What is Ana going to do?" Some students said, "Draw twelve squares." Carlos responded with another question, "What other information is there? Not any old squares. What else?"

This may seem like a typical interaction, but it is, in fact, quite unusual. Before Ana presented to the class, Carlos asked the students to reflect on what she was going to do, with details. This is a metacognitive invitation to reflect on the mathematical process. When the students were not specific enough, Carlos encouraged them to think in a more detailed way. As Ana prepared to draw her rectangle, Carlos then shared an important invitation with the class. He told all of the other students that they had a specific role to play when Ana was presenting, which was to be respectful and to consider questions that they could ask her, so that they could all "have a good conversation about the work."

In this request, Carlos was asking students to reflect on their own behavior and ways of interacting with the ideas of the speaker. He was working to bring metacognition to the moments of listening. Notably he pointed out that even if Ana has a correct solution, the other students have a role to play in considering and discussing Ana's work by posing good questions. Teachers all know that when students are presenting to a class, it is common for other students to switch off and lose focus. Carlos averts this by giving students a role, asking them to think of questions for the presenter.

As Ana started to draw her rectangle, Carlos asked her to speak out loud and explain her thought processes, another important metacognitive request.

Ana drew a rectangle and declared, "This is twelve." Carlos immediately responded, "How do you know it's twelve?" This is an important question. When we ask people, *How do you know that an answer is correct?* it invites them to reason about their answer, which is a critical mathematical act. Carlos carefully and explicitly works to expand students' ideas of what it means to be mathematical and to add reasoning to their toolboxes so that they can be successful with any future mathematics problem. When Ana shared that she counted to twelve, Carlos responded, saying, "That is a nice strategy. Does anyone have another strategy?" This statement both valued Ana's strategy and communicated to students that there are different approaches and that he, as a teacher, valued those different strategies.

Later in the lesson, Alfonso drew a rectangle for the class that was more than twelve squares, then announced that it was wrong and drew a cross through it (fig. 2.1).

Carlos responded, "How do you know it is wrong? Why did you cross it out?" As Alfonso faltered at the front of the room, staring nervously at his incorrect solution, Carlos jumped in with affirming words: "You did a lot of perfect math there, and I want to hear what was going through your brain." At first Alonso said "never mind" in a dejected way, but Carlos persisted with his communication that Alfonso's work was important and if he could explain what was wrong, he would move everybody forward.

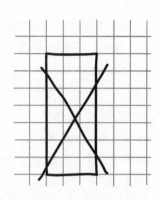

2.1 Alfonso's incorrect "12" rectangle

In these moments Carlos commu-nicated something valuable to all the students in the room. He treated the mistake in the same way he would treat correct work, showing interest in the mathemat-ics, asking why it was wrong, investigating the mathematical process, and giving it value. He also took special care to praise Alfonso, telling him he had done "a lot of perfect math." This way of valuing the student and his mathematical thinking was an important moment in the lesson. Alfonso responded to the encouragement Carlos gave and explained, correctly, think-ing in threes, and drew a rectangle that had an area of twelve (fig. 2.2).

Later Carlos shared with me that both of the two students that presented, who were the first stu-dents in the class to present, had statements of special educational needs, and he was thrilled that they had been so courageous. He said that in those first moments, his main job as a teacher was to

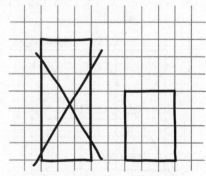

2.2 Alfonso's revised rectangle

protect them and value their mathematical thinking, whether it was correct or incorrect.

The next moments in the class were equally important as Carlos invited the students to share their strategies for finding twelve squares. A few students offered that they multiplied to get twelve, which would be the end of the conversation in most mathematics classes. But Carlos came back to the answer of multiplication, asking, "Why?" A few students shared that multiplication worked because it was faster or because it gave the answer. This did not satisfy Carlos, as he wanted the students to focus on the mathematical process. In the discussion, Carlos asked the class "Why?" seven times, until a student eventually explained that "each row is a group of 3, and there are four rows." In these moments Carlos shared that he valued the mathematical process, and understanding why it works, and that an important role for the students in the class was to pursue deeper meaning. The act of asking someone *Why?* is one that we can all practice in interactions with others.

Later, Hector illustrated the process of multiplication, circling the parts of the rectangle that corresponded to the numbers (fig. 2.3), and Carlos praised him, saying that kind of explaining and coding helped people see each other's thinking.

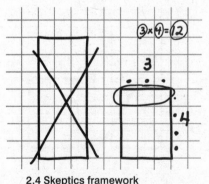

2.4 Skeptics framework

Carlos then expanded on the importance of this coding, saying that technical writers use this kind of careful coding and "technical writers are people who make a lot of money."

Carlos ended the class period

with these encouraging words: "We are going to do a ton of problems with area. So if you understood that, you're in great shape. If you didn't get it yet, you're not in terrible shape. You don't need to panic or worry, but you do need to keep working on area problems when they come up."

In these moments, Carlos was asking the students to reflect back on the mathematics they were working on—finding area using multiplicative thinking—in a simple but meaningful communication, which also signaled to students that they needed to learn this content, but if they did not know it yet, that was also OK, as they would receive further opportunities.

In this short extract of teaching, Carlos invited students to move from additive to multiplicative thinking, which is important mathematical content. He did this through inviting students into eight different types of metacognitive thinking. He asked students to listen respectfully, to talk out loud, to think about what someone was going to present, to consider different strategies, to understand and value errors, to think about why methods work, to notice color coding and "technical drawing," and to think back on what they learned. Throughout the mathematical discussion, Carlos stands at the back of the room to communicate to the students that they should exchange ideas with each other, not just with Carlos.

Engaging students or yourself in metacognitive thinking is a key part of developing a mathematical mindset, becoming open to reflective thinking and diversity in thought. If you are a parent, a teacher, or someone who works on problems with others, you can engage in similar conversations, encouraging others to be metacognitive. A key invitation to offer others, or yourself, is the question *Why?* When people consider why they chose a method,

or why a method or rule works, or why they chose a business strategy, they enter a deeper, more reflective zone of thought.

For teachers, whole-class discussions are particularly valuable in both encouraging and highlighting metacognitive acts and strategies for the whole class.

Carlos's sixth graders went on to reach very high achievements, with students from the class going on to meaningful jobs and different colleges, including Stanford and the University of California. One student even took her place on the school board as an eleventh grader and went on to become a member of the California State Board of Education.

ENCOURAGE METACOGNITION THROUGH EIGHT MATHEMATICAL STRATEGIES

Perhaps the most important tools in a metacognitive person's toolbox are the strategies they use when they approach a mathematics problem. These strategies are often what separate successful mathematics learners, and people generally, from less successful ones, as highlighted by a range of research studies.[20] I think of many of these strategies as superpowers, as they enable people to be successful but are not well known or well used by learners or problem solvers.

1. Take a Step Back

The first strategy, useful in all areas of learning and work, is the approach of stepping back from a problem, considering what this problem is asking from you. This may sound obvious, but most people read mathematics questions and problems and think they should be able to answer them immediately, or give up. When his students call Carlos Cabana for help, unsure of how to start a problem, he asks them to say the problem aloud.

"What is the problem asking you? Say it out loud."

When he requests this and students say the problem out loud, they often immediately follow up with, "Oh, I know what to do." It is as if repeating a question out loud unlocks mathematical thinking. This is one of the reasons it is so important that students work together in groups, as they receive opportunities to describe and talk about problems with each other.

A good follow-up question is, "What is this question about?" This is broader and invoking of metacognitive thought.

I am aware that a common strategy shared in mathematics education is to encourage students to look for key words when they are working on mathematics word problems. For example, teachers tell students to look for the word *of* in mathematics problems and to use that as a cue to divide, or to look for the word *more* and add the numbers. I am not at all convinced that key words are productive, and this is why: we really need students to understand the meaning of problems and to consider, "What am I being asked here?" When they look for key words, they are not taking in the bigger picture and considering the overall meaning; they are doing the opposite, hoping that the key word will prompt them to use a particular rule or method. Not only does this approach result in incorrect answers, but it also encourages students to miss the meaning of questions altogether.

2. Draw the Problem

This is a strategy I use in every single mathematics question I work on, and I cannot say enough about the value of this approach. As I will share in chapter 5, researchers have found that the brain activity that separates mathematicians from other academics comes from visual areas of the brain—and this is true whatever the mathematical content.[21] Asking people to draw a numerical question will cause activity in the visual parts of the brain, as well as stimulate important connectivity between numerical and visual areas. It will also give people a different way to approach and understand any problem, which is so important.

3. Find a New Approach

A third strategy is to ask people to think about a different approach to a problem. I have found this strategy to be particularly effective with high-achieving students, many of whom have only ever used one approach to solve problems. This is also a helpful approach to use in classrooms when students work at different rates. If students finish work before others, I do not give them different work until I have asked them if they can think of other, different ways to solve the problem they have just completed.

This is an important way to introduce students to thinking with mathematical diversity.

4. Reflect on "Why?"

In the sixth-grade classroom, Carlos asked the students no fewer than seven times why multiplying works before a student went beyond saying "it works" to explaining the mathematical process and the logic behind it. Knowing why something works is critical to students' understanding, particularly for girls and women, who have been consistently found to desire this depth of understanding at greater rates than boys and men.[22] Knowing why is good for all genders, but women and girls often turn away from mathematics when they are not given access to this depth of understanding.

5. Simplify

Make problems easier to understand, calculate, or see. As chapter 6 will reveal in depth, the action that separates high from low achievers when working on number problems is that of changing the problems.[23] For example, when asked to add 19 + 6, some students instead add 20 + 5. This may seem like an obvious strategy, but I have found that many low-achieving students believe that it is somehow "not allowed" to change the problem that is given to them. The approach of changing numbers or shapes helps people become more flexible in their problem-solving.

6. Conjecture

A sixth strategy to teach is to invite students to come up with their own conjectures. A conjecture in mathematics is an idea that has not yet been proved, is still at the idea stage. In science we would call this a hypothesis. It is very interesting to me that students in schools all know the meaning of a hypothesis, but most have never heard of a conjecture. This says something about the problems in our mathematics education system. The widespread focus on rules leaves many students missing the value in something as playful as a conjecture. In my teaching, I tell students that part of their role in my class is to come up with their own conjectures, and this often shifts their whole mathematical perspective.

7. Become a Skeptic

A seventh strategy, which pairs with students' work providing conjectures, is to invite students to reason and to take on the role of a skeptic. I share with students that it is very important in mathematics to share reasons—to explain why they have chosen methods, the logical connections between them, and why they work. This is called reasoning and it is the essence of mathematics. When mathematicians publish their research, it is full of mathematical reasoning, as that is the method by which mathematicians prove mathematical ideas. Adults who can share their reasoning are also more effective in the workplace.[24] I tell students that there are three levels of reasoning: at the lowest level, you can usually convince yourself of something; it is slightly harder to convince a friend; and the highest level of reasoning is being able to convince a skeptic (fig. 2.4). Then I tell them to be skeptics!

I encourage this metacognitive approach—the skeptics framework—because I have found it to be incredibly effective

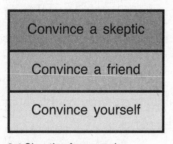

2.4 Skeptics framework

at changing classroom discussions. Before I share this framework, discussions are often directed at me, but when I encourage students to be skeptical with each other, this changes.

When I tell the middle school students in our youcubed camps that they need to be skeptical, they embrace the idea immediately and willingly. I was worried in our first camps that the class conversations would often be directed at the teacher, with students answering our questions but not having conversations with each other. When I asked the students to be skeptical, this immediately changed. I still remember a student called Josh proving a conjecture that Matt, another student, had given. Matt had shared the conjecture that the sum of any two odd numbers will always be an even number. I asked the class if anyone could prove this to me. Josh volunteered and approached the board at the front of the room.

Josh started his reasoning with an example, saying 1 + 2 = 3, and 3 is odd, but 1 + 1 = 2, and 2 is even, adding cheekily, "And this works for everything. The end!" The first responses from students, who had learned to be skeptical, were, "Why does that

work?" and "Prove it to us!" Josh accepted the challenge, saying happily, "You want me to puroooove it?!" He then shared another example: "I will add 201, which is odd, to 1,103, which is also odd, and end up with 1,304, which is even." The students continued their skepticism, asking him *why* this works. Josh added to his proof, saying, "Because you can divide 1,304 by 2 and it will split into two even numbers. Bam!"

In these moments, the students practiced skepticism, and Josh responded with further reasoning. This link between skepticism and reasoning is powerful and encourages young people to know that their role in mathematics learning is to engage in sense-making and reasoning, one of the most important ways of being mathematical, valuing mathematical diversity, and gaining access to understanding. As students reason more, other students are given greater access to understanding.

8. Try a Smaller Case

A final mathematical strategy is to ask students to solve a problem by working with a smaller case. I share this strategy with all my students, as I usually find that they have never been taught it, yet I consider it a mathematical superpower.

For example, if you were asked to work out how many squares are on an 8 by 8 chessboard (fig. 2.5) (the answer is not 64!), it would be helpful to first determine the number of squares on a 2 by 2, a 3 by 3, and a 4 by 4 chessboard (fig. 2.6).

2.5 Chessboard

This work with a smaller case will allow you to see the underlying patterns much more clearly.

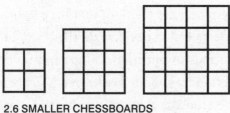

2.6 SMALLER CHESSBOARDS

Left, Can you find 5 squares?; *middle,* Can you find 14 squares?; *right,* Can you find 30 squares?

This approach can be used in all areas of mathematics, and many other forms of knowledge. For example, if you were asked to divide 2 by $5/6$—a question that causes many people to feel anxious and break out in a sweat—a good strategy is to ask yourself, "What is 2 divided by $1/6$?" and build up from the smaller case.

These eight strategies will help you in any mathematics question or mathematical work you need to solve. They are not very well known or used in mathematics classrooms, yet they convey great power to learners. A handout sharing these ideas is available to download from Mathish.org.

ENCOURAGE METACOGNITION
THROUGH JOURNALING

I am a big fan of giving learners journals in which they reflect on their mathematics learning journeys. These are not the exercise or workbooks typically given out in math class, where students record answers; they are instead open spaces for free thought and reflection. It is not only students who benefit from having journals in which to set out their thinking and their reflections; we all benefit from having spaces for our reflective thoughts. I do not go anywhere without my own journal to record my thoughts, ideas, and plans.

I prefer journals with pages that are blank or that have lightly dotted squares, so that students can think outside the lines, literally. We give students journals in our math camps and invite them to write down any useful ideas about mathematics or their own learning. We also give students time at the start of class to decorate their journals, so they feel ownership of them. Occasionally we collect the journals and give students comments on their work. We leave our comments on Post-it notes so that students do not feel that their own space for thinking has been taken over by teacher evaluations.

BUILD A REFLECTIVE AND GROWTH MINDSET

Studies have shown that reflecting on your own learning—an important part of being metacognitive—boosts achievement.[25] Knowing how to self-reflect often makes the difference between being an effective and ineffective learner; therefore, teachers and

parents should encourage this practice at all times and in differ-ent ways. The table on page 48 shares options for useful reflec-tion questions that can be given to students as part of lessons, as classroom exit tickets, or for consideration at home. The list is intended to provide choices, so that students can be prompted to reflect on one or more statements at different times. As a parent, you can ask your children questions like this as part of your everyday conversations. Or you can go further by asking children to put the questions in their journals, so that they can regularly reflect on them.

When teachers I have worked with replace typical homework questions, which are often not meaningful, with the request to think back on the lesson (reflect) at home, their students report that it increases their mathematical understanding.[26] This prac-tice is so effective because students are given opportunities to reflect on their own knowledge and understanding, which is in-credibly valuable. Teachers I worked with a few years ago would choose one or two questions for each homework assignment. The following are comments that students shared after their

	What mathematical ideas/concepts did you learn today?
	How is the idea you learned today related to others you have learned?
	What opportunities did you get to struggle? How did that feel?
	How could you use the mathematical concept in your life?
	What different strategies or approaches to the problems were helpful to you?
	Are there areas that you do not understand and would like more opportunities to learn?

Can you write your own problem for someone else to try to solve?

2.7 Reflection ideas

homework changed from narrow practice questions to reflection questions:

> I feel like the homework questions help me reflect on what I learned from the day. If I do not quite remember something, then it gets me a chance to look back into my composition book.

> Having the reflection questions does help me a lot. I can see what I need to work on, and what I am doing good on.

> I think the way we do our homework is very helpful. When you spend more time reflecting about what we learned, and less time doing more math, you learn more.[27]

Reflection also plays an important role in helping students develop a growth mindset, which is another part of metacognitive thinking. Many teachers have asked me how to identify whether their students have a growth mindset or not. Some have tried mindset surveys but not found them to be useful; across the world, most students know the "right" or expected answers to give in mindset surveys. What is more important than answers to surveys is the ways students behave when work is difficult or they make mistakes. To help both teachers and students become aware of their own mindsets and learn important metacognitive strategies, my team at Stanford developed a mindset rubric, shown in figure 2.8.

Teachers have told me that this rubric has helped their students develop mathematical strategies and a growth mindset. Some teachers give the rubric to students at the start and at the end of a course, collecting responses at both times to see if they change.

	Beliefs: I believe in myself, and I know I can learn anything, as I have unlimited potential. I know that my brain is flexible, and it is developing, strengthening, and/or connecting pathways all the time.
	Struggle: When I find work difficult and struggle, I keep going, knowing that I am developing my brain. I am not afraid to take risks, try something new, and get it wrong.
	Strategies: If I try a method or approach that does not work, I try a different approach and think about the problem in new ways. I like to investigate ideas, search for patterns, and think in different ways: visually, verbally, and physically, as well as numerically.
	Connections: I am curious about other people's ideas and their different ways of thinking. I ask questions about what I am learning to reach new understandings.
	Reflection: I think reflection is a valuable learning practice. When I get a lot of feedback, even if it looks overwhelming, I know it will be helpful and I use strategies to incorporate it into my own work.

2.8 Mindset rubric

ENCOURAGE METACOGNITION THROUGH GROUP WORK: TEACH STUDENTS TO RESPECT EACH OTHER'S IDEAS

A critical time for students to learn ways to engage metacognitively is when they work with other students in a group. I have been studying equitable group work for many years now,[28] while using the strategies with my teaching students at Stanford and in our summer camps.[29] As I mentioned earlier, I witnessed some of the most impressive examples of group work during the years I studied Railside School, watching students from grade 9, with few or no metacognitive strategies, to grade 12, by which time they had become highly effective communicators and problem solvers. This did not happen by chance; the teachers were intentional about creating effective group members who respected and helped each other, and themselves, learn.[30]

The teachers created effective group work using the strategies of complex instruction, an approach that is designed to make group work equal and reduce status differences among students.[31] The Railside teachers spent the first ten weeks of every class focusing on respectful student interactions in groups. This dedicated time really paid off in the students' interactions for the rest of their years in the school. During that ten-week period, the teachers gave the students especially "group-worthy" tasks, activities for which different group members are needed. When the students worked together, the teachers highlighted and shared respectful interactions, creating a useful feedback loop.

One way in which complex instruction encourages good group work and metacognition is the allocation of roles that students enact in their groups. The authors of complex instruction recommend that students work in groups of four with four roles.[32] Over the years, and after teaching the roles to teachers in England, I have adapted the method slightly and added a fifth role, set out below.

Giving each student a role in the group has many benefits and

Limitless roles, adapted from complex instruction

Includer: Ensure everyone is included in the group work. Ask your group to read through the task together before you begin. Ask everyone, "How do you see it? How do you think about the ideas? Does everyone understand what they have to do?" Keep your group together and work to open ways for people to understand. Make sure everyone's ideas are heard. Check with others: "Are we ready to move on?" Make sure your poster shows everyone's ideas.	**Connector:** Remind your team to think in different ways, to search for connections, and to find reasons for each mathematical statement. Some different ways the team may think include visually, with words, with numbers, with graphs, with movement, or with modeling. Make sure your team's poster is well organized, uses color coding, arrows, and other math tools to show connections. Think together, "How do we want to show that idea? How do we want to highlight that connection?"
Synthesizer: Pay attention to time, space, and resources. How long should your team spend on different aspects of the task? Think about where the group is and where they need to go in the remainder of the time. Make sure there is space for deep and creative thinking. Be responsible for the resources your team needs. Consider: "Are there any supplies we need that can help us visualize this more? Are there any resources that might help us solve this?"	**Questioner:** Your role is to be curious and encourage curiosity in your group. Ask lots of big and open questions and encourage others to ask creative questions too. Push the group to think in divergent ways. Be a skeptic and encourage others to be skeptical—pushing others to give reasoning for their ideas. At some point be ready to join the teacher for a huddle.
Spy: Once during the group time, you can spy on one other group. Be stealthy, and do not take photos!	

can even be adapted for use with children in the home. One benefit is that all students feel included and have something specific to do. The roles also encourage metacognitive thought, pushing students to ask why and to think deeply and strategically. When the teachers at Railside invited students to take and fulfill their roles, they used a strategy I have found to be highly effective in my own teaching: a participation quiz.[33] This involves giving the students a task to work on in groups, but before they start working, they are given a set of valued mathematical approaches. The Railside teachers shared the following list, a set of mathe-

matical behaviors that, again, invite students into metacognitive thoughts and actions:

Mathematical Ways of Working

During the participation quiz, I will be looking for:

- Recognizing and describing patterns
- Justifying thinking using multiple representations
- Making connections between different approaches and representations
- Using words, arrows, numbers, and color coding to communicate ideas clearly
- Explaining ideas clearly to team members and the teacher
- Asking questions to understand the thinking of other team members
- Asking questions that push the group to go deeper
- Organizing a presentation so that people outside the group can understand your group's thinking

The teachers also shared this mantra: *No one is good at all of these things, but everyone is good at something. You will need all of your group members to be successful at today's task.*

In addition to the list of mathematical ways of working, the teachers shared a list of good group behaviors:

Good Group Behaviors

- Leaning in and working in the middle of the table
- Giving equal airtime to all members
- Sticking together
- Listening to each other
- Asking each other lots of questions
- Following your team roles

Teachers walked around the class, taking note of good (or bad) group behaviors, writing down actual words students used on the whiteboards or on paper. At the end of the lesson, students got feedback, and sometimes grades, for their group behavior. When I teach my undergraduate and graduate classes at Stanford, I enact participation quizzes (though I call them participation boosts) whenever I see group work becoming unequal. This almost always happens at some point, with some students directing mathematical conversations and others choosing not to participate or being left out. Unequal group work changes when I enact a participation boost, and the students become more aware of their behavior. This metacognitive practice is an important step in helping students to appreciate mathematical diversity in their groups. Sharing with them what is valued in their work— prompting them to think about the way they are learning, to develop metacognitive strategies, and to take responsibility for their learning journey—leads to long-term success.

Another strategy the Railside teachers used was giving group quizzes, in which they collected the work of just one group member (randomly chosen) and everyone in the group would get that person's grade for that piece of work. This very strongly communicated that everyone in the group was responsible for the learning of everyone else.

ENCOURAGE METACOGNITION
THROUGH ASSESSMENT

A chapter on metacognition would not be complete without some discussion of the powerful role that assessment can play in encouraging metacognitive, self-aware people. I have written elsewhere about the importance of changing students' perspectives from a performance culture to a learning culture.[34] A performance culture is one characterized by the lack of useful information on ways to improve, with only test scores and grades being used to communicate students' performance. A learning culture is one in which a teacher's feedback informs their learning and helps them know ways to improve. On the occasions I have been in schools that created a learning culture, the students told me how grateful they were to receive information on their own learning. (Chapter 7 will share an exemplary case.) I am never surprised to hear this, as the students have been invited to take charge of their own lives and to be responsible for their own progress. As assessment-for-learning leaders describe it, such students have been invited into the guild.[35]

The very best way to communicate to students their learning journeys in mathematics is the use of mathematical rubrics. These show what is to be learned and where students are in the learning process. Sometimes teachers add comments that are extremely valuable for students, and sometimes students are asked to self-reflect on their own learning. Our youcubed website shares an example of a school that changes to a learning culture, to encourage the development of growth mindsets, by changing assessment from tests to rubrics with feedback; in chapter 7, we'll see an example of a teacher using rubrics to give students signposts to guide their learning journeys.[36]

The act of setting out steps to reach a student's learning goal and evaluating progress toward the goal does not have to be an assessment-only approach. For example, in my last book I told the story of Milly and her amazing teacher Nancy Qushair, who is the head of a mathematics department in an International Baccalaureate school.[37] Milly came to Nancy's class feeling that she could not be a "math person" as others worked faster than her. She described herself as "dumb." To help Milly, Nancy suggested they focus on one topic. Milly chose integers, so Nancy gave her lots of integer problems using different visual representations and providing feedback on all her work, which helps students know where they are in their learning. By the end of the year, Milly was a different person. She wrote Nancy a letter saying that seeing ideas visually and knowing why they worked, as well as how, had changed everything for her. Milly's transformation came about because Nancy introduced Milly to mathematical diversity, she gave her clear goals, and she evaluated her progress toward the goals. As I was writing this book, Nancy contacted me to say that Milly had been sitting in her college mathematics class at the University of Oregon when the mathematics professor played a video from my last book, *Limitless Mind*.[38] The video features Milly talking about her own change.

This was a beautiful, full-circle moment of Milly being helped by mathematical diversity and then playing a role in helping others.

Another example of providing a mathematics learning goal with feedback on progress came from a friend of mine helping her third-grade daughter learn multiplication math facts. Kristina was concerned that her daughter's school approach was entirely numerical and memorization based, so she gave her daughter, Abby, a new notebook and invited her to write a chapter for each multiplication number—that is, a chapter on her 2 times table, another on her 3 times table, and so on. She asked Abby to take a week or so on each chapter and include the following:

1. A visual showing one or more of the math facts
2. A noticing about the number pattern that she saw
3. A real-world example, giving meaning to the numbers

It could have been daunting for the third grader to consider doing this work for all twelve sets of number facts, and she could easily have gone astray with mathematical thinking, so Kristina made it fun and set clear goals. At the end of each chapter, Kristina would review the work with Abby, give her feedback, and ask her questions; she would also play a fun number game with her that would highlight the numbers. Kristina thought she might have to provide an extrinsic reward, such as a treat for each completed chapter, but she was thrilled that she did not need to: Abby liked having twelve areas to work on, one per week—it was a manageable goal, with feedback, and the number games were fun. Abby also felt that the work was helping her when it was math time in her classroom.

Figure 2.9 shows a page from Abby's notebook:

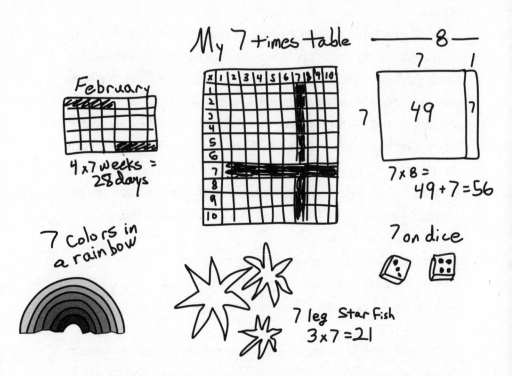

2.9 Abby's notebook highlighting the number 7

Many educators, leaders, and parents are aware that people approach learning differently, but they do not realize that the effective ways of learning and problem-solving can be taught. Most teachers devote their time to teaching their content area, assuming that students know how to learn. In fact, students not only do not know the best learning approaches, but they often have learned approaches that are counterproductive, especially if they have been subjected to a harsh performance culture and narrow mathematics.[39] Importantly, all of this changes when students are taught metacognitive approaches to learning, which teach them to be open-minded and inquisitive about diverse mathematical ideas.

In this chapter I have shared ways to encourage effective metacognitive learners: class discussions, a range of powerful mathematical strategies, journaling, reflecting, respectful group work, and assessment that gives feedback to students. When we teach metacognition through these ideas, we teach something that is generative for people's lives. These ideas are important, as when we teach someone to be open to diverse ideas, to ask questions, to think deeply, and to reflect, they will learn how to learn, and they will benefit more deeply from every different idea they meet in their lives.

3

VALUING STRUGGLE

As we progress through this book together, experiencing the beauty of mathematical diversity, and the playfulness of math-*ish* ideas, there is a mindset that is important for everyone to embrace. This mindset changes not only our learning but the way we interact with the world. It comes from the important science of mistakes, struggle, and challenge.

Extensive research provides evidence that people who have a growth mindset are more effective in their lives.[1] But what does it mean to have a growth mindset? Many think it means that you know you can learn anything, and that trying is good. Both of these beliefs are important, but I see the defining quality of *mindset* as a set of beliefs that surround the times when we struggle, make mistakes, and experience hard times in our lives. A key feature of a growth mindset, which protects people when things go wrong and helps folks persist through difficult problems, is a changed reaction to times of struggle, and mistakes. Neuroscientific studies have shown that whereas people with a fixed mindset view mistakes as evidence of their own weakness, people with a growth mindset view mistakes as opportunities to learn. In electroencephalography studies, people with a growth mindset, compared with those with a fixed mindset, show better error correction following feedback, greater neural markers of attention to feedback, and lower neural markers of emotional distress due to errors.[2] In other words, they experience mistakes very differently from people with a fixed mindset. When people are encouraged to develop a growth mindset, they start to approach mistakes in a positive way.[3]

The implications of this research are profound—imagine that in every difficult situation you face, you feel strengthened, you feel encouraged to learn, you pay better attention to feedback, and you have more effective brain responses that increase your learning. It is no surprise that multiple studies show that people with a growth mindset are higher achieving in all levels of education.[4]

I am sharing this mind shift, which allows us all to welcome challenging ideas and situations, at this early point in this book so that readers can approach the ideas of mathematical diversity and math-ishness with an open and generous mind. This is especially important for anyone who has ever experienced math trauma or adversity, which—let's face it—is many of us! I have written about this topic in previous books as it is central to my research and work. In this chapter, we will explore the latest evidence and ideas around embracing struggle to improve our understanding, learning, and lives as we set out on our mathematical journey together.

LEARN TO LOVE STRUGGLE

In the past ten years or so, I have received hundreds of emails from parents asking for advice about their child's learning of math. It is impossible for me to reply to them all, although I try. One recent email caught my attention. It came from a parent of a Stanford undergraduate, imploring me to help his daughter. Julie had enrolled at Stanford ten years prior, but due to a serious illness had taken many months off school. He told me that she hated math, had significant math anxiety, and had one statistics course standing in the way of her graduation with a degree in English literature. It was still early in the quarter, and Julie was finding the material "incomprehensible," and, given her disability, the course was taking her to the "verge of a complete physical collapse," according to her father. I met with Julie and was immediately impressed by her intellectual curiosity and her excitement for ideas in English. I was equally struck by her deep sadness as she talked about math. She had not had good experiences with math in high school, and she was convinced that she could never pass the stats class.

I arranged for one of my doctoral students, Margie Hahn, to tutor Julie. At first, Margie told me that Julie was failing every class test, but the pair persisted. Together, they embraced the struggle. Notably they started reviewing class tests each week, going through the questions that Julie got wrong. Margie would help Julie see the ideas differently and find other approaches to the questions than she had been shown in her lectures. Gradually, Julie started to improve. At the end of the quarter, I received a lovely text from Margie telling me that Julie had just achieved an A– on her final exam. Julie—and her dad—were thrilled. Julie graduated from Stanford and is now pursuing doctoral studies.

When Julie focused on her mistakes and thought about a different approach to problems, she became unlocked, and her learning flourished. This was made possible by an important teaching act from her professor—giving students their tests back for reconsideration and revision. This does not happen enough in teaching, yet mistakes are one of the most important sites for any learning.

I teach an undergraduate class at Stanford called How to Learn Math. Each year I show the students a video of a middle school math lesson in Japan. I tell them that the video comes from a study that explored the nature of teaching in different countries. The study had chosen Japan to observe as it is a country of very high mathematics achievement.[5] My students are always enthralled by the lesson.

The Japanese class is made up of approximately forty students with mixed prior achievement. In the beginning of the lesson, the teacher introduces a visual investigation—students are shown a piece of land divided between two people.

They are given the task of making the line that divides the land straight, without changing the amount of land each person owns. The question has no numbers and it is challenging, as the land is unevenly shaped, lacking ninety degree angles that would make dividing it equally much more simple; my undergraduates usually work on it before they watch the video, but few of them come up with a solution. The teacher in Japan gives the students time to work on the problem in groups. Interestingly the teacher also gives students a choice: he tells them that as they work on the investigation, they can either discuss the problem with their friends, go to the teacher for help, or get hint cards.

As the Japanese class investigates the visual problem, the teacher moves around the room preparing the different groups to present some of their ideas to the rest of the class. The class is a vibrant space, with some students standing in groups, some sitting, and lots of movement, conversation, and laughter around the room.

The teacher smiles and laughs as he talks with the students, showing the positive relationship he has developed with them. Almost all the students are smiling as they work with their friends in groups. It is during this time that an interaction occurs that shocks and fascinates my undergraduate students. The teacher asks a group of students to share the mistake they had made with the rest of the class. The students look at him quizzically and ask, "You want us to share what we did wrong?" "Yes," he says, "the mistake is what's important. If people can do it from the beginning, then they don't need to come to school." This is a strong statement—the teacher communicates to the students that he expects them to make mistakes, that they are a natural part of learning, and that he values the mistake enough to share it with all the students, as it is a source of productive learning for everybody.

The lesson continues with different students sharing mistakes and solutions, smiling and laughing as they do so. The teacher then extends the investigation, sticking some premade shapes onto the board, posing further investigative questions, asking students to use their knowledge of lines and angles to adapt shapes, and encouraging students to come up with a variety of different approaches.

This lesson is shocking to my students, for many reasons. One is the investigative nature of the lesson, especially as many of my students believe that high-achieving countries such as Japan and China drill students in facts and rules. The second reason is an idea we will explore in depth in chapter 5—the teacher's provision of visual models, including the cardboard shapes he had premade. The third is his encouragement and the sharing of mistakes. The researchers who studied the teaching approaches in different countries also discovered something concerning for those of us in the US: in Japan, students spend 44 percent of their time "inventing, thinking, and struggling with underlying concepts," but in the US, students engage in this behavior less than 1 percent of the time.[6]

Steve Olson is an award-winning science writer who studied students in different countries taking the International Mathematical Olympiad. He drew this conclusion about the role of struggle and mathematical diversity in classrooms in Japan:

Teachers *want* their students to struggle with problems, because they believe that's how students come to really understand mathematical concepts. Schools do not group students into different ability levels, because the differences between students are seen as a resource that can broaden the discussion of how to solve a problem. Not all students will learn the same thing from a lesson . . . , but each student will learn more by having to struggle with the problem than by being force-fed a simple, predigested procedure.[7]

In chapter 1, I discussed the villain of the math brain. The opposite of this is the strong and replicated fact that all students are on a growth journey and their brains are strengthening, connecting, and expanding all the time.[8] In addition, times of struggle, mistakes, and failure are not signs that a person is weak or bad at a subject; conversely, they are signs that incredible brain activity is happening.

When I teach students, whether middle and high school students in camps or undergraduates at Stanford, I tell them, "I am giving you difficult work as I want you to struggle. I want you to work on something challenging, so that you can get a great brain workout." This is freeing for students, who become more willing to persist knowing that the time of difficulty is a productive one. I tell them that the work feels hard because their brains are working so hard. Something important goes with this message: the provision of low-floor, high-ceiling problems—problems that anyone can access but that reach to high levels. The problems I share also have multiple access points, including visuals that students can readily make sense of. Students work on problems with challenge, knowing that they can be successful, which is important. In recent years I have heard of teachers telling students to struggle and work alone without help, but the problems they give students are typical, narrow math questions. Students get frustrated and do not know how to access the questions, and the learning is not productive. What makes the time of struggle and challenge productive for students is the provision of mathematics tasks that embody mathematical diversity and "ish"—later chapters in this book will share many examples of these tasks.

Research in mindset shows that when we believe we can learn anything and we engage in struggle, we are more likely to benefit from those times of struggle than when we do not believe in ourselves or the importance of struggle.[9] Part of the reason we have such low mathematics achievement nationwide is that so many teachers, and students, believe that struggle is a sign of weakness. If we could change teachers', parents', and students' responses to struggle, we would unlock higher mathematics achievement at all levels.

I was making one of my online courses when I first heard Carol Dweck talk about the moment that she decided to start her research on mindset—which would turn out to be one of the most influential ideas in education across the world.[10] She told me that she had been interviewing some young children, giving them tasks to work on, and saw that many of the students withdrew when faced with a difficult task, but one young boy reacted completely differently. When he was given a challenging task, he exclaimed, with glee, that he loved a challenge, and he jumped in to work on it. This young boy, unbeknownst to him at the time, played an important role in worldwide education. His exclamation caused Dweck to realize that the way we approach tasks might change the way we learn. After decades of research by her and her colleagues, this theory was proved and came to be known as *mindset*.[11]

When students have a growth mindset, as opposed to a fixed mindset, they believe that they can learn anything, and they view mistakes and challenges as opportunities for learning. This mental approach not only bolsters them when learning is difficult but also protects them from damaging stereotypes, and it encourages persistence—all of which lead to higher achievement.[12] The evidence for the impact of a change in students' mindsets from fixed to growth is overwhelming: research has shown that mindset interventions can raise achievement,[13] improve health,[14] help students dial back on aggression,[15] and reduce racial disparities[16] in the classroom.

Dweck witnessed firsthand an important difference between most of the students she worked with that day and the boy who was not only undeterred but excited by difficult work. The challenge this offers—to parents and educators—is how to develop this kind of thinking, and excitement to face challenge, in our students and in ourselves. Over the past ten or so years, many research studies have investigated whether a mindset intervention prompts a more productive approach to learning. The evidence is complex, showing that some students are changed when they learn about brain growth and the importance of times of struggle. But more recent research is showing that for wide-scale and lasting impact, these messages need to be specific to mathematics and delivered through a change in teaching, not just in the delivery of different ideas.[17] In other words, what we really need is the development of mindset cultures in classrooms and workplaces. The teaching approach that helps develop a mindset culture starts with perhaps the most important teaching and learning condition of all—the creation of environments in which students are encouraged to struggle.

The neuroscientists I have worked with over the past few years study different aspects of our brains, using contrasting techniques, but all of those I have met have been clear about one neuroscientific finding: the most productive times for our brains are when we are struggling and making mistakes. Bestselling author of *The Talent Code* Daniel Coyle studied those who are most successful in their different fields and concluded that they are all people who have learned by "working at the edge of their understanding"—making mistakes, correcting them, moving on and making more mistakes.[18] This confirms my own experience: students who are prepared to struggle and work at the edge of their understanding are usually the students who learn more than any others.

Ellie, a student who was willing to work at the edge of her understanding, attended the first youcubed summer camp. If you had watched the class that Cathy Williams and I taught, you probably would have noticed Ellie, as she was very willing to share her thinking and answers, which were usually wrong. An observer may have regarded Ellie as one of the lowest-achieving students in our class, and it is true that she did come into the class with one of the lowest pretest scores. But I really appreciated Ellie sharing her mistakes, as it gave all the students important thinking opportunities. Ellie would argue for her mistaken approach in a determined way, eventually correcting her thinking and moving on. She was working at the edge of her understanding.[19] When we analyzed the students' achievement gains, Ellie had improved more than any of the other eighty-two students, becoming one of the highest achievers in the group.[20] There was a lot of evidence that Ellie really heard the messages we gave about struggle and mindset, and this changed her approach to mathematics. Years later, when she was a senior in high school, Ellie wrote to me and asked for my help as she had decided to write her senior school project on mindset.

Steve Strogatz is an applied mathematician at Cornell who conducts research on "small world networks."[21] This concept draws from the idea that we are all connected by six degrees of separation at most, and we live in a small world. Steve conducted research with one of his students at the time, Duncan Watts, which showed that many systems in the world—from power grids to the neural networks of worms, to collaboration among Hollywood actors—do not operate randomly but as clustered networks with characteristic path lengths. Their paper communicating research on small-world networks is one of the top 100 most-cited scientific studies in the world.[22]

There are many things I appreciate about Steve's communications about mathematics learning, none more than his descriptions of his own mathematical struggles. In a beautiful podcast interview, economist Steve Levitt interviews mathematician Steve Strogatz and talks with him about his work and his mathematics journey.[23] During this interview, Steve shares that he "bumbled" his way through his math degree, getting the lowest grades of any subject he took, because the courses did not involve intuition or visuals—in other words, the mathematics did not have any diversity; it was one-dimensional. He also shares the time he fell in love with the feeling of struggle.

Steve had a memorable experience in high school when his teacher gave his class a problem to solve and shared that no previous student had ever been able to solve it. The teacher also told the class that he, himself, an MIT graduate, had not been able to solve it. This caught young Steve's attention. He started working on the problem, and after an hour or two he could not get it. He said he worked on it day after day, and then the days turned into weeks and then months. After six months he developed a correct proof. His teacher was so pleased he shared it with the head teacher.

In those months Steve became hooked on the stubborn pursuit of understanding through struggle, which he called the "fight." He loved the feeling so much that he started creating difficult problems for himself, asking questions of this broad and amazing subject we call mathematics.

Steve's journey into becoming one of the world's leading mathematicians refutes the myth, common in the education system, that struggle is a sign of weakness and is only experienced by "low-level" thinkers. Struggle was a key part of Steve's progress as a mathematics learner. Another dangerous myth is that the most successful people do not struggle. This is completely wrong—in fact, the most successful people are those who are comfortable with struggle and, like Steve, learn to seek out the feeling of struggle, knowing that it is an essential part of accomplishment.

This is important for all of us to remember: we should embrace or even seek out opportunities to do work that pushes us to the edge of our understandings, as it is on that edge that the greatest knowledge can be discovered; it is where creativity is found and important discoveries can be made. Often, when we venture out onto the edge of places where we are unsure, lack knowledge, or have uncertainty, that is where we achieve the greatest accomplishments. We do not achieve much in life by playing it safe or giving in to our inner negative voices and fears.

In four different experimental studies, with students learning different content at different ages, researchers have compared the approach used in most math classrooms—teaching methods, then students practice the methods in questions—with a different approach.[24] In the contrasting condition, teachers give the students questions and tasks *before* they teach the methods they need to solve them. The students are invited to use their intuition to discuss possible ways forward. All the studies show that this teaching approach brings about higher outcomes, and the researchers conclude that this happens because students get a greater opportunity to struggle—to think about and draw from the knowledge they have already developed.[25] When students learn new methods *after* they have struggled to work out a way forward, their brains are primed to learn the new material.

One of the studies, an investigation of students learning calculus at Harvard University, compared students working in different approaches.[26] In one of the approaches, students received a lecture before they worked on problems. In the other approach,

students worked on problems before learning about the methods they needed. The study was designed well, so that groups of students experienced both contrasting approaches at different times. Interestingly, the study found that students learned more from struggling first, working on problems before they learned methods. However, it also found that students *thought* that they had learned more from listening to a lecture first. The clarity of the lecture had given students the illusion of learning, and the experience of struggle had made them feel bad. The researchers concluded that one of the reasons the students believed that the struggle condition was not as effective was because they had never been taught the value of struggle. Despite the students' thinking the approach was less effective when they struggled, both sets of students learned more when they were given problems to consider before they learned the methods that would allow them to go forward.

We have extensive evidence that the best times for our brains are times when we are challenged and struggling.[27] This evidence has profound implications for the ways we live our lives, and it is the opposite of the messages many of us have received from schools and other institutions.

"BOARD THE STRUGGLE BUS"

Anders Ericsson was a Swiss psychologist and professor at Florida State University who was often recognized as a world expert in the development of expertise. Ericsson described the important learning route to developing expertise as one of trying, failing, revising your approach, and trying again, over and over.[28] It is no wonder this country has such low math achievement, as we rarely encourage students into this important approach. Instead, classrooms and homes focus on correctness, praising students for getting questions that involve little opportunity for struggle correct. Ken Robinson, an internationally acclaimed leader in education and creativity, famously shared that it is impossible to do anything creative without making mistakes.[29] I would say that it is impossible to work on any valuable mathematics that is appropriately challenging without making mistakes. So how do we help students develop comfort with this important and valuable process—one that is incredibly helpful for learning, understanding, and life? This is not a minor task, as so many people feel bad each time they make a mistake and thus try to avoid difficult challenges. Here are some strategies that I and other educators have found to be the most effective in shifting students' thinking and their approach to learning and life.

A NEW PATHWAY FORMS

PATHWAYS CONNECT

PATHWAYS STRENGTHEN

3.1 Brains forming, connecting, and strengthening pathways

Share Neuroscience

I always start the courses I teach by sharing with students that our brains are growing, connecting, and strengthening pathways all the time, as can be seen in figure 3.1. There is no such thing as a "math brain"; our brains are changing all the time.[30] I want my students to struggle and make mistakes. When we are struggling, it is a really important time for our brains: they are forming, connecting, and strengthening pathways.

Start a Class or Conversation—or Family Meal—with Discussion of Struggle

An interesting and important conversation to have with people is one where you share the value of struggle and ask others how they feel about it. Ask them how often they experience struggle and how it feels in the moment. If you can, share times when you have struggled with math, or share Steve Strogatz's story, described earlier in this chapter. Carol Dweck encourages people to acknowledge that we all have times of fixed mindset thinking, when we think we cannot do something. She even encourages us to give a name to our fixed mindsets. When you discuss struggle with your children, students, or colleagues, ask them to think about situations when they develop fixed mindset thinking, and brainstorm ways to change that thinking into a recognition that struggle is a positive sign that they are doing something important.

We know that the most important place to be for the development of understanding and brain connectivity is on the edge of our understanding.[31] One important practice is communicating how important that place is, and how we should want to be walking on the edge.

I also like to ask students to share with me when they think they are on the edge, so that we can consider it together. Invite students or children to share with you: When does it happen? How does it feel? Make the idea of working at the edge an important one that students carry in their own minds and that they are proud to experience. Invite them to draw themselves on an edge and carry the image with them in their books.

Share Metaphors for the Value of Struggle

A few years ago, I was contacted by Alina Schlaier, a graphic designer in Germany. She told me her eight-year-old daughter, Greta, had significant math anxiety until she worked on creative, visual mathematics, and she learned about the value of struggle and mindset. Alina herself had been successful in math; she had excellent teachers in school who showed her various ways to think about mathematics. When her daughter declared she hated math in first grade, Alina was shocked and knew she had to do something. She found the challenging but accessible, diverse problems we had created and shared on youcubed.org and in books, and gave them to Greta.[32] She said her daughter loved them and would start asking her family to play math games at dinnertime. I was thrilled to learn of Greta's transformation. I was also interested to hear about Alina's. Even though Alina had been successful in math, she said it had always been a subject of numbers. As she started working on math problems with her daughter, she started seeing math concepts with "new eyes." She told me that she had never before thought about how symmetry divides positive and negative numbers or how helpful it was to see multiplication, visually, as area. She told me she now solves mathematics problems better at work and in other areas of her life.

Many app developers had contacted me over the years to work with them on math apps, but in all cases the apps they were developing only accepted one answer. I had often seen students become disillusioned when they did not input the answer the app "wanted" in the form they were looking for. When I talked with Alina, she showed me the beautiful designs her team could create and we shared the goal that any app should value the different ways students think. We decided to develop an app together, and because we wanted to help students feel good about struggling, we called the app Struggly.[33] In Struggly, people experience mathematics through visual tasks and games, their different ways of thinking are valued, and they earn badges for struggling and thinking in different ways (fig. 3.2). The badges incorporate various metaphors for highlighting the process of struggling.

One of the metaphors we share with our Struggly users is to think of struggle as a cloud that has come down over their minds. We tell them their minds may feel foggy and they may feel that they cannot think clearly.

We tell students it is important to stay in that moment of struggle, to feel it and appreciate it, and to keep thinking. Eventually brain sparks will fire through the clouds of struggle.

Depth Over Speed   **Challenge & Mistakes**

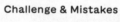

Making Brain Connections

3.2 Struggly badges

Those brain sparks come about when we start having ideas. Struggly provides a range of ideas for pulling through struggle, such as:

- talking about the problem,
- drawing the problem,
- writing about it,
- taking a break, such as a short walk,
- approaching it differently, and
- thinking outside the box.

One metaphor that has been hugely generative for many people comes from James Nottingham: the pit of struggle.[34] Some teachers encourage students to make a pit together, such as the one shown in figure 3.3, made by Jen Schaefer's students.

As students are going into the pit, they use words such as "This question is too hard" and "I can't do it," but as they are coming out of the pit, they have reframed these thoughts, using words such as "I can do this," "School is meant for trying things out and making mistakes," and "I just need another way." They also show a path across the pit, with the words "Danger, do not take this road." Jen Schaefer, a teacher in Canada who shares the pit with her students, tells them that she could take their hand and walk or jump across the pit with them, but that would not help them as it would take away their opportunity to struggle. Jen works to make sure her students feel comfortable with times of struggle, providing them with accessible tasks that allow this to happen.

Sometimes Jen's students come to her and say, "Ms. Schae-

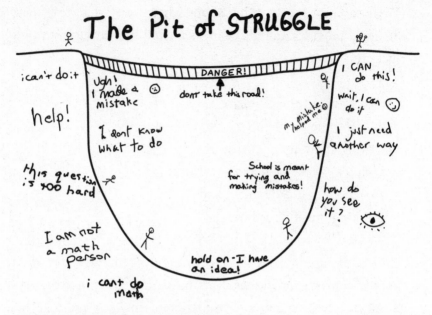

3.3 Student depiction of their Pit of Struggle

fer, I am really in the pit!" Her response to that is "Excellent! What tools do you need?" I love this response for two reasons. First, Jen values students for being in the pit, and second, she does not dive in and structure the problem to make it easier for students to avoid struggle; she asks them what tools they would find useful.

When I was visiting a school recently, I learned of an elementary teacher who invites her students to construct an interpretive dance about the learning pit, and the teacher shared that she worked to infuse the idea into all of her teaching.

These different metaphors—learning edges, pits of struggle, and stormy clouds—have all been incredibly meaningful for students of different ages, as well as for adults wanting to live with a growth mindset, valuing and learning from times of struggle.

Give Challenging Work but with Plenty of Access Points

As students learn mathematics, they need work to be challenging so they get an opportunity to struggle. But math questions and tasks that offer this opportunity in the right way have a particular set of qualities that are important. It is not helpful to present people with narrow questions that they cannot answer and have no way to think about. The type of questions that prompt struggle, and comfort with struggle, are very different. One quality they all possess is known as a "low floor and high ceiling." The example that opened this book, in which students are invited to share where they see the extra squares, is a good example of this quality (fig. 3.4). Everyone can share where they see extra squares (low floor), but the task extends to high-level algebraic thinking about quadratic functions (high ceiling).

I have found that problems that encourage struggle and comfort with struggle, as well as the beauty of mathematical diversity, have a range of different features, which are shared below.

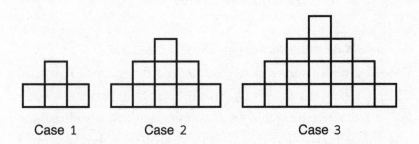

Case 1   Case 2   Case 3

3.4 Growing pattern of squares

Diverse problems that invite students to the edge of their understanding should:

	Be low floor and high ceiling—the low floor means anyone can access them; the high ceiling means they extend to high levels
	Include visual or physical thinking
	Be possible to see and solve in different ways
	Invite ideas and discussions
	Include reference to the world

These problems happen to be the most engaging and exciting for learners, and I highly recommend them to you. We will meet many of them in the remainder of this book.

Celebrate When Students Struggle and Make Mistakes

When I give difficult work to students and they express how hard it is, I tell them that it feels like that because their brains are working so hard—forming and strengthening pathways and making new connections—and that is exactly what they should want. I tell them that it is an important moment and to value the feeling of work being hard. In my own teaching, I also, like the Japanese teacher, put mistakes on the board so that the whole class can benefit from the mistakes and see their value. Jen Schaefer asks her students to write a math reflection in which they share their favorite mistakes or their biggest struggle or aha moments. Jen sent me some of her students' reflections on aha moments, and I was struck by the huge mathematical value of students talking about complex ideas such as subtracting integers, understanding the unit in the addition of fractions, and rearranging algebraic expressions.

Lisa van den Munckhof, a teacher I met when visiting the Senpaq'cin School, a First Nation school in Kelowna, Canada, told me that when teachers or students make mistakes in class, they are given five high fives from others in the class. I love this idea, and when I visited Lisa's class, I could see that a mistakes-friendly culture had been established. I will share more about my enjoyable visit to the school in chapter 6.

For parents, it is important to value mistakes in the home, which can be difficult at times, particularly when your children break your favorite ornament, spill a drink over your work papers, or become overexuberant and knock over their younger sibling. I know how hard it is to react calmly and supportively in such moments; I have had to work on this myself in my own parenting. But we know that students' ideas about themselves, their mindsets, and the importance of struggle all start in the home. A study by a group of mindset researchers found that by the time they were three years old, children had developed different mindsets depending on the types of praise parents gave.[35] So next time your child makes a mistake, work hard at communicating the helpful message: Mistakes are a part of life and an important learning opportunity.

Use Growth Rather than Fixed Praise

One of Carol Dweck's studies powerfully showed the impact of different forms of praise. The researchers conducted an experiment in which two groups worked on a difficult task. One group was praised for working hard; the other group was told that they were really smart. Both groups were then offered a choice between a challenging or an easy task. Ninety percent of those who were told they were smart chose the easy task, whereas the majority of the students praised for working hard chose the challenging task.[36] This study shows how a small change in the way we praise people can have an immediate effect. We know that when we praise others by telling them they are smart, they at first feel good about the idea, but when they mess up later or fail, as everyone does, they start to think, "Maybe I am not so smart." It is fine to praise children, but praise what they have done, and the process they have undertaken. Tell them you love their creative thinking, or their cool approach to a problem, rather than using fixed labels. Those praised for being smart chose an easy task in the study because they were afraid of losing the "smart" label that they had been given. This is a common response to fixed praise: it leads to vulnerability, wherein people work to protect their "smart" image, and takes away their comfort to ask questions and to struggle.[37]

Change the Way You Assess

In chapter 7, I will share different ideas for giving people useful assessment feedback. If we are encouraging students to feel good about mistakes and times of struggle, we cannot penalize them every time they make a mistake. If we do, this sends a strong contradictory message, of which students will be keenly aware. Of course, on a final test, we all know that it is important for students to make as few mistakes as possible, but assessments given throughout a course or year of teaching are very different; they are opportunities to encourage struggle and mistakes.[38]

Share Famous or Important Mistakes

Marc Petrie is a middle school teacher in Santa Ana, California.[39] Marc shares a video each week with his students that shows an example of someone exhibiting a growth mindset. I love that idea and think it could be extended to include mistakes, as the world is filled with examples of mistakes that may have seemed problematic at the time but eventually became valuable. One of my favorites to share is the story of the solving of Fermat's Last Theorem.

Fermat's Last Theorem is named after a French mathematician, Pierre de Fermat, who made a bold claim in the 1600s that the equation $a^n + b^n = c^n$ has no whole number solutions when n (the exponent) is greater than 2. When n is 2, we can create a number sentence that is true: $3^2 + 4^2 = 5^2$, but Fermat stated that no set of numbers would ever work when n is greater than 2. He also scribbled in the margins of his work that he had a "marvelous" proof that it would never work, but he did not provide the proof. This set mathematicians on a quest to find the proof for hundreds of years. It was more than 350 years later that a proof was found by the shy English mathematician Andrew Wiles. Notably, Wiles first encountered the problem of Fermat's Last Theorem when he was a ten-year-old boy in Cambridge, and he describes the moment when he learned about it, saying,

> It looked so simple, and yet all the great mathematicians in history could not solve it. Here was a problem that I, as a ten-year-old, could understand and I knew from that moment that I would never let it go, I had to solve it.[40]

Years later, after Wiles earned a PhD and became a mathematician, he started working on the problem in earnest. Wiles explored and searched for patterns, for years, working to construct a valid proof. One day, seven years later, he emerged from his study in his house and announced to his wife that he had a proof.

The venue for sharing the proof was the Isaac Newton Institute in Cambridge. Rumor had gotten out that Fermat's Last Theorem may have been solved, and the room was packed with more than two hundred mathematicians and journalists. Wiles presented his work over three different lectures, and when he finished the room erupted into applause. Over the next few weeks, it transpired that there was a mistake in his work, and Wiles went back to his study to work for a few more months before he came up with the complete and accurate proof.

This is an interesting story in itself, but one of the parts that I love to highlight is the fact that the mistaken theories that had been put forward as proofs of Fermat's Last Theorem have now generated new fields of mathematics—including parts of algebra. As author Peter Brown has described, "Out of the ruins of these failures rose deep theories that opened up vast new areas of mathematics."[41]

The middle school students that we teach in our summer camps are really struck by the history of Fermat's Last Theorem. First, they are amazed that someone worked on a math problem for over seven years—a math problem that they can understand. This changes their ideas about the time they "should" spend on problems. They are also impressed that the mistakes generated new and important areas of mathematics; the story helps to underscore the value of mistakes in human thought and discovery.

Crumple Paper

Around 2013, during the time that online courses, or MOOCs, which stands for "massive, open, online courses" were exploding onto the scene, I had begun working with the thought leader Sebastian Thrun, a computer science professor at Stanford, who had recently invented self-driving cars. Sebastian had just started a new company to create online courses called Udacity. My work advising Udacity led to my creating my own MOOC for teachers of mathematics that was released as a Stanford on-line course.[42] I was not sure if anyone would take the course but was pleasantly surprised when thirty thousand people took it during that first summer. Inside my first MOOC, I shared the value of mistakes and asked course participants to design an activity that would help students see the value in mistakes. One of my favorites was sent in by a teacher. She suggested that students crumple a piece of paper with the passion and frustration they feel when they make a mistake and throw the paper at the front board.

Students are then invited to unfold their piece of paper and color in the lines, with all the lines representing the brain growing and making connections.

Choose Your Favorite Mistake

A popular practice among teachers that I appreciate is the choosing of a "favorite mistake" to share each day. This has the benefit of showing appreciation for mistakes but also enables consideration of the mathematical principles that underpin the ideas being learned in class or at home. This is helpful as the domain of understanding around any mathematics problem always extends beyond the problem itself and is important to consider.

For example, when students are asked to add the fractions $2/3 + 1/4$, there will always be some students who get the answer $3/7$. This is a valuable mistake that is completely worthy of class discussion. If I were leading this class, I would start by saying, "This is an interesting example, as Naj has added the top numbers to get 3 and the bottom numbers to get 7. That approach is important for us all to think about. But others in the class got the answer $11/12$. In mathematics, there are questions that have more than one right answer. Is this one of them? *(I wait for students to answer.)* If we agree it is not, then we have the fun challenge of working out which answer is correct, and why."[43]

In class discussions like this, the authority in the room shifts from the teacher, who could easily give the correct answer, to the mathematics, as students are asked to reason their way to a solution. I love it when classes give me different answers to questions, as that tells me we have an interesting problem, with lots to talk about. When I open the conversation to students to share their reasoning, I also invite students to come to the board at the front of the room and defend the example they believe is correct, (hopefully) adding visuals. When we reach a conclusion, together, that an answer is correct, I always value the role of the mistaken answer in the development of understanding.

A similar exercise that I really like is to give students another student's work, which is actually a problem that you have made that includes an important mistake, and ask your students to give feedback to the student. It is always good to spend time considering why approaches to problems do or do not work. When students make mistakes in mathematics, there is almost always some logic to their thinking, and it is valuable to highlight and consider that logic.

These are all examples of ways to discuss mistakes and nonstandard examples, making mathematics much more interesting and diverse.

Share Videos and Articles

Our website, youcubed.org, shares a range of resources to help students feel good about struggle and mistakes. These include videos[44] and news articles from Science News.[45]

Scientists and educators have known about the value of students struggling and making mistakes for generations. Long before neuroscientists showed the value of such times for our brains, Jean Piaget (1896–1980), the Swiss psychologist, talked about the value of students being in a state of disequilibrium, a type of "cognitive imbalance" that prompts us to modify our models of learning and move to a state of equilibrium.[46] Lev Vygotsky (1896–1934), another giant of psychology and learning, focused on what he called the "zone of proximal development"—the space between what students can do without assistance and what they can do with assistance from an experienced guide.[47] Both of these psychologists knew that the times students were in disequilibrium, or needing adult assistance, were the most important times in their learning. Now we know that times when students are struggling and working at the edge of their understanding are the most useful times for brain activity and development. Despite this wealth of evidence, students everywhere feel bad about struggling and making mistakes, which negatively affects their learning going forward.

As a society—not only in our schools—we need to recalibrate our cultural fear of being wrong and struggling, as they are both times of great importance and value—for the brain's development of knowledge and creativity.

Perhaps the place to start helping learners become comfortable with struggle and mistakes is in helping adults become comfortable with the same. This often starts with working to eradicate negative self-talk, something we all suffer from at times. Becoming comfortable with uncertainty, mistakes, and struggle allows adults to model being comfortable in these times with other adults and with learners; demonstrating comfort is an important part of the process. Wherever you are in this journey, I hope that this chapter has been helpful in sharing ideas and resources to help you create mistakes-friendly environments for your learners and help you embrace times of challenge and mistakes yourself. Learning is a process, not an outcome, and the times of struggle are when the most dynamic opportunities occur.

We have now reached the time for me to share my new ideas for learning mathematics, which I hope you will embrace with the open, positive mindset that these first chapters have set out.

4

MATHEMATICS IN THE WORLD

My goal in writing this book is to share the idea of math-ish, and to celebrate mathematical diversity. Math-*ish*, and the concept of "ish," as you will read later in the chapter, is a diverse approach to understanding mathematics as it exists in everyday life for many different kinds of people. The concept of mathematical diversity encompasses the value of cultural diversity and difference in people, and diversity in approaching mathematics, appreciating different ways of seeing and thinking. Chapter 1 shared that both forms of diversity, especially when they are brought together, help encourage effective collaboration, problem-solving, and high achievement.[1] In the youcubed summer camps, my team and I have taught students who were diverse, and the teaching was successful because the mathematics approach valued and elevated student differences, through our respect for their varied ways of seeing, thinking, and problem-solving.[2] Student diversity enriched the mathematical diversity, and mathematical diversity supported the student experience.

The most compelling and large-scale research to consider the impact of racial diversity comes from Sean Reardon, one of my faculty colleagues at Stanford, who investigates educational inequality. Sean and his colleagues have looked at educational opportunity, drawing from an impressively large dataset:

eleven years of achievement data from more than fifty million students in over ten thousand districts. This data reveals an invaluable takeaway: racial segregation of students is associated with achievement gaps starting in third grade and expanding through eighth grade.[3] Schools are not as racially segregated today as they were sixty years ago, but racial and economic segregation has increased in the past thirty years. In 2022, the average Black student attended a school with 32 percent more Black and Latine students than the average white student.[4] When schools and communities are not diverse, everyone loses out, especially as many schools and communities are underfunded, leading to severe and indefensible opportunity gaps that disproportionately effect students of color and students from low-income communities. Reardon's research suggests that, if equality is an important goal for us as a society, we need to reconsider the organization of our schools. Reardon's research shows the value of more racially diverse schools in countering inequality, a result that is not surprising to me. Through all my years working with schools in the UK and the US, I have found that the most effective are those with considerable diversity of all kinds—racial, cultural, social, and more. People benefit from opening their minds and appreciating the diverse contributions of different cultures and people.[5] With careful teaching, young people develop respect for each other that crosses racial, social, and cultural divides—one of the most important goals of education.[6] Diversity enhances life, and it enhances mathematics.

A few years ago, I received a phone call inviting me to help create opportunities for a more diverse mathematics to enter the

K–12 school system. It came from someone outside of my usual professional circles—someone who was not in education or in mathematics. The caller was Steve Levitt, an economist from the University of Chicago, who became famous for his book *Freakonomics*.[7] He invited me onto the *Freakonomics* podcast, which is usually hosted by his Freakonomics colleague, Stephen Dubner. What brought Levitt into the conversation was his frustration at the mathematics homework his high school–age children were being asked to do and, more generally, the content of the mathematics they were being taught. He recognized that it was the same mathematics he had learned in school, yet his own work conducting economics research involved mathematics and mathematical tools that had "nothing to do with what my kids are learning." This disconnect prompted Levitt to host a special edition of the show entitled "America's Math Curriculum Doesn't Add Up."[8] The episode opened with one of his high school daughters reading aloud in a droll voice:

> Rationalize the denominator in the equation: 3 over the square root of x minus 7. Find the imaginary zeros of the equation: f of x equals 4x to the fourth plus 35x squared minus 9.

A lively conversation followed, involving myself, economist and former teacher Sally Sadoff, research analyst Daphne Worchenko, and president of the College Board, David Coleman. Much was said about the disconnect between the mathematics that *should* be taught and the mathematics that *is* taught. Coleman reported that the College Board had surveyed college professors of math and other subjects and high school math teachers, asking them all what mathematics is most needed for college courses. He reported that the difference in the answers given by the two groups would "break your heart":

> The college teachers say, "Very few things matter and matter a lot." The high school teachers say, "Everything matters." Think of the stress of that. They must do everything, or they are betraying their kids, which forces them to race through the curriculum lest their kids are not ready. What the college teachers say but is not heard is, if your students can do these core set of things, we can do the rest. But if those are shaky and they're merely faintly aware of them and aware of a lot of other mathematics, we're stuck.[9]

These are strong and important words. For the college teachers, *few things matter, and they matter a lot*. But the high school teachers believe that everything must be covered. This is not surprising, as the high school teachers have been told—by states, districts, curriculum standards, textbooks, and tests—that they should teach everything that is set out in the standards. (In chapter 6, I will attempt to help with the problem that all K–12 teachers, and some parents, face: ways to teach content meaningfully and with the diversity students need when there is so much content to teach.) You may be wondering which are the few things that matter? What are the areas that college professors believe are critical for different college courses and for students' mathematical futures more generally?

LEARNING AND TEACHING WHAT MATTERS

The first core math concept David describes as "humble": arithmetic—the four operations and fractions, an area of mathematics I describe as number sense. The second area is data analysis and problem-solving, with concepts such as rates, ratio, and proportion, allowing people to see how quantities relate to each other; I call this data literacy. The third is linear equations, an area of mathematics that describes how things in life relate to each other that is used in many disciplines. These three areas not only are needed for college courses but also happen to be the areas of mathematical thought that are most needed in life and work. Each can be encountered in a narrow way—as a set of rules, methods, and procedures—or in diverse and ish ways that highlight their potential, their beauty, and their applicability to our lives. This chapter is going to start our consideration of these three areas. What is interesting about them? And how can we encounter them in diverse and ish ways?

Number Sense

Numbers are so important, but they have a negative reputation in society at large. Most people do not know that numbers have an innate coolness; this is not surprising, as most people never get to see or experience what is cool about them. In fact, many people think of numbers and arithmetic as cold, remote objects, irrelevant to their lives. The curricula set out by most countries and states engender this idea. King's College London professor Stephen Ball describes the "curriculum of the dead" as curriculum and standards that do not recognize the role of people in the creation of knowledge, appearing instead to be handed down "by the unassailable judgment of genera-tions," and that do not include space for students' own experi-ences or representations.[10] The lists of mathematical methods set out in national standards and curriculum certainly fit that description. When those standards are translated into narrow textbook questions by publishing companies, students come to view numbers as cold hard facts lacking any cultural diversity, or diversity of any kind.

Numbers: A History

One way that I challenge the lifeless, person-less version of num-bers that students experience is through sharing the rich cultural history of their development. Few students get to learn about the history of numbers, but if they did, they would learn that the first quantitative records in the world came from the Brazilian Amazon.[11] Paintings created by Indigenous artists more than ten thousand years ago show "x" marks, counting days, moons, and other cycles. Archeologists have found numerous pieces of evidence that the ancient people of Brazil were paying attention to quantities, now thought to be the first prehistoric numerals.

In a small area of central Africa, a bone was found that has fascinated historians over the generations. The Ishango bone has deep mathematical significance; it is considered to be twenty millennia old and has a set of markings that reveal awareness of prime numbers and the decimal system. These first indications of the important parts of our number system came from what is now called the Democratic Republic of Congo.[12]

I like to ask students to look at a diagram of the bone that comes from a great book of activities from around the world by Claudia Zaslavsky (fig. 4.1). I ask questions such as *Why do you*

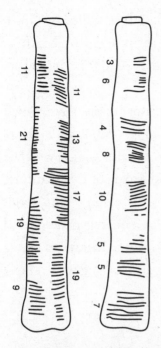

4.1 Two sides of the Ishango bone
C. Zaslavsky, Math Games and
Activities from Around the World

think these numbers are on the bone? What do you think the bone was used for? This allows them to deepen their appreciation for the diverse nature of our mathematical history, and it opens up their understanding of ish numbers in the world.

Sumer is the earliest known civilization that lived in the region of what was called Mesopotamia, which is now "the land between the rivers" in Iraq. Later the Babylonians lived in the same region, and it is the Sumerians and Babylonians who are credited as being the first users of algebra.[13] The majority of clay tablets that have been recovered there, showing algebraic markings, date from 1800 to 1600 BCE. The word *algebra* comes from the Arabic: al-jabr, meaning the reunion of broken parts. Moving through time, the number system we know and use in the West comes from the Arabic system which, in turn, came from the Indian system. It was scholars in India who invented the number zero.

One of the first uses of numbers is shown in the recording of time. People wonder why our system of recording time is in units of 12 or 24 (hours), and the answer is that it comes from the ancient Egyptians, who developed sundials more than three thousand years ago. The decision to divide the day into 12 units came from the shade they could see on sundials. They saw 10 units from sunrise to sunset and then added a unit for dawn and another for twilight. The system was formally codified by Greek astronomers in the Hellenistic period, around 323 BCE. These decisions have remained through time and are the reason that we have 24 hours in each day.

These are just a few examples of the history of numbers, which are often not shared with students. I find the examples fascinating; they show the rich cultural history of mathematics, which is beautifully diverse, spans the world, and is part of every culture.

I think of mathematics as a lens that we can all lay on the world; when we do that, we see so much more, noticing patterns and interesting relationships. Importantly, numbers themselves have a rich cultural history, worthy of knowing and sharing with others.

The Allure of Patterns

Numbers have always fascinated me, and I am sure this comes from my early experiences of numbers being introduced to me as a visual and physical playground. I spent hours with a set of Cuisenaire rods that my mother brought home for me, which seem simple but are incredibly powerful (fig. 4.2). Invented by Belgian teacher Georges Cuisenaire, each of ten colored rods represents a number from 1 to 10. Cuisenaire noticed children learning musical relationships by playing keys on a keyboard. He wanted them to have a similar sensory experience that would give them insights into number relationships.[14]

If you are a parent of young children, I encourage you to buy a set of Cuisenaire rods and just set them out. Your children will instinctively start to play with them, order them, and investigate patterns.

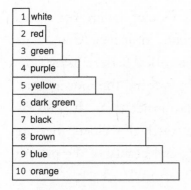

4.2 Cuisenaire rods

As a young child I loved those colored rods and spent many hours ordering them in different ways, investigating number patterns. When I got older and had a calculator, I would spend a lot of my time in school (when I was meant to be listening to teachers) punching in numbers and changing them with different operations to see what happened to them. I kept a notebook, a practice I have maintained all my life. My childhood notebooks were often filled with number patterns that I would continue to investigate for days. I hope readers do not take from this that I am some sort of math genius, drawn to numbers at an early age. If there is anything that can be gleaned from this story, it is that I learned to play with numbers from a young age; I was able to experience them visually, physically, and flexibly; and through this I developed a curiosity for and fascination with numbers.

Now, as a professor at Stanford, I have managed to reconnect with my early love: one of the activities that I invite undergraduates to work on involves building patterns with Cuisenaire rods. Like many of the activities I value and that we share on our youcubed website, the activity appears to be simple—anyone can access it, including young children—but it extends to very high levels that challenge undergraduates. One of my favorite videos shows one of my undergraduate students, Yasmeena Khan, sharing a visual proof of a complex mathematical pattern using Cuisenaire rods. She shows the proof by moving the rods around, highlighting the mathematical pattern inside them. It is no exaggeration to say that when people watch this video, they are shocked and moved by the intricacy and beauty of the pattern that unfolds before them.

People really enjoy watching the video, and they ask me if they can watch it again and show their students. I love to see people's reactions to Yasmeena's visual proof, as what they show in those moments is mathematical appreciation, and their appreciation is for something very pure—a beautiful, visual number pattern that highlights a deeper meaning than the numbers can convey alone.

Many ask what Yasmeena is doing now; they guess she may have become a creator at a high-tech company. Yasmeena works at a global problem-solving firm. She shares that she continues to be grateful for my undergraduate class, as it changed her

mindset and approach to mathematics, which helped her take "multiple advanced math classes (linear algebra, multivariable calculus, probability and stats)" at Stanford, enabling her to go on to further success in the world.

There is great value in sharing with people mathematical activities that relate to the world, but I also know that people of all ages are fascinated by the patterns inside and between numbers. David Coleman, CEO of the College Board, highlights the importance of number sense for students entering college courses, and it is my firm belief that the best foundation in number sense that we can give to anyone starts with an invitation to play with numbers, to explore patterns and to approach them in ish ways. When numbers become visual, the experience is even more meaningful.

Brent Yorgey is a math and CS professor at Hendrix College. When I first encountered his beautiful representations of numbers, I was enchanted.[15]

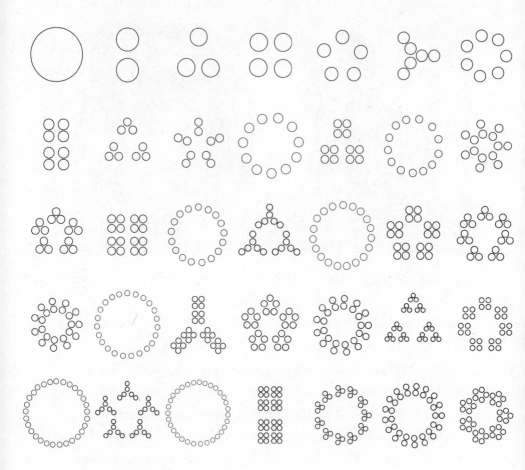

4.3 Brent Yorgey's number visuals

My team and I often give the visual in figure 4.3 to students and teachers and ask them, first, to write each number next to its corresponding visual, so that they can *see* the number and get visual insights into the way it is made up—its factors and its relationships to other numbers. Then we ask people to look for interesting patterns. When we do this activity in classrooms, students excitedly share the different patterns they see. Some of them are shown in figures 4.4–4.7:

- The multiples of 3 all have a similar structure that shows their 3-ness—they all include triangles.

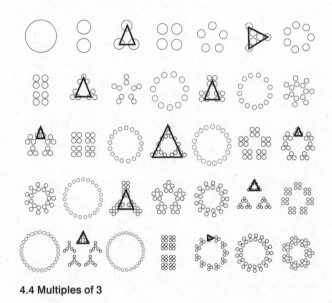

4.4 Multiples of 3

- The multiples of 7 are all in seven-sided shapes.

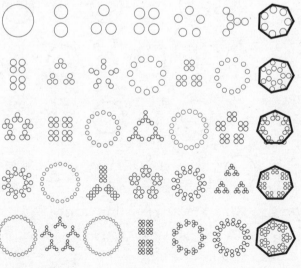

4.5 Multiples of 7

- The multiples of 6 are all in the same shape, like this:

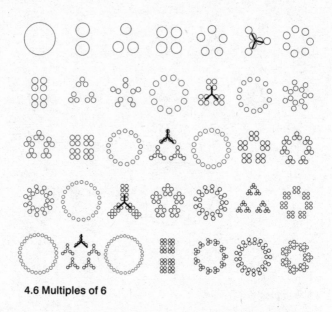

4.6 Multiples of 6

- The prime numbers (except 2) are shown as circles.

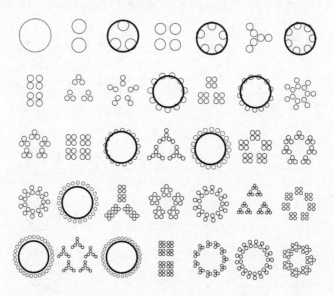

4.7 Prime numbers

I particularly appreciated one teacher I worked with in Arizona named Randy, who noticed that the prime numbers are shown as circles. He made a passionate argument that the number 2 is also a circle, which he showed as two circles in orbit (fig. 4.8).

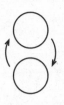

4.8 The number 2

At that moment Randy was engaging in mathematical flexibility, an important creative act we will explore in chapter 5.

I once shared this representation of numbers with a middle school class and asked the students to explore patterns. A few days later, when I visited the school again, a parent approached me. She asked me, with an excited and animated tone, what I had done with the students in class, as her daughter, who had hated math and did not think she could ever like it, had changed her mind. I have experienced this response many times before; it comes when people have a realization that they can "see" and play with numbers—and that mathematics is a creative domain. When young people realize that numbers are not remote hard facts but cool and lovable characters, everything changes in their learning.

Giving people of all ages an opportunity to simply play with numbers and number patterns has intrinsic value, beyond the standards it checks off in any curriculum. Once people have played with interesting number patterns, they start to see numbers and mathematics differently.

Numbers can always be seen visually, and the visuals add deeper meanings. I remember the time I was working with a group of elementary teachers when one exclaimed with joy that she had not known that square numbers have this name because they can be drawn as a square:

1	4	9	16	25

4.9 Square numbers

This visual representation of square numbers is so much more meaningful than a number with a squared exponent next to it. When we share that 4 is a square number because it can be drawn as a square (a 2 x 2 square) and 9 is the next square number that can be drawn as a square (3 x 3), people are often shocked and fascinated. I also shared with the teacher that day that square numbers come from the addition of odd numbers, which can be seen in figure 4.10. The numerical version of this is as follows:

1 + 3 = 4
4 + 5 = 9
9 + 7 = 16, and so on

I also shared triangle numbers with her that day— numbers that can be arranged as triangles (fig. 4.11).

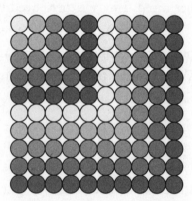

4.10 Addition of consecutive odd numbers

1	3	6	10	15

4.11 Triangle numbers

The teacher was totally enchanted by the visual representations of numbers, which she had only ever known as symbols.

I have been teaching undergraduates at Stanford for many years, and despite their many mathematical achievements, most of them have never heard of triangle numbers. You may not think this is a great loss for these young students, many of whom will change the world through their work in education, in charities, in companies they found and run, and in new inventions they create, but triangle numbers are needed in many areas of mathematics, including probability, algebraic functions, and more. The students had learned these areas without the benefits of this interesting representation of numbers. This is, of course, just a symptom of a bigger problem—mathematics had been introduced to students as a symbolic numerical subject, void of mathematical diversity and joy. When we invite students to know and explore triangle numbers, the visual representations of numbers, and the place of numbers in our history and culture, we invite them into the world of number patterns and flexibility, an important place to be.

Math-ish

Numbers are everywhere in the world and we all use them, in some form, every day of our lives. But there is something noteworthy about our everyday use of numbers that differs from the ways we use and learn numbers in school. When we use numbers in our lives and workplaces, they are nearly always imprecise estimates, what I call ish numbers. To some people the idea of ish numbers is heresy, as they believe that numbers have to be accurate, precise, and correct at all times. But ish numbers turn out to be the numbers we most need in our lives, and I believe they could transform people's approach to mathematics if they were present in their learning journeys. Here are some questions that would typically be answered with ish numbers:

- How old are you?
- How much of the moon can we see tonight?
- Can I have half of that cookie?
- How long is the drive to the airport?
- How warm is it outside?
- How big is the United States?
- How long is London Bridge?
- How much flour do I use in this recipe?
- How much paint do I buy to paint the wall?

These are just a few examples of ish numbers in the world. Shapes are always ish shapes, as there are no perfect circles, triangles, or rectangles. Examples of ish shapes, such as sugar cubes, kites, concentric circles, and pyramids, shows us just how ish-like many common shapes are.

Ish numbers and shapes are everywhere; we see them every

day. I share them here not only because they are curious and interesting but also because they are important. When educators in the UK wanted to improve mathematics teaching, they arranged for a committee, led by esteemed mathematics educator Sir Wilfred Cockcroft, to consider the mathematics used in the workplace. One of the most critical areas highlighted by the committee was estimation, which they described in this way:

> Industry and commerce rely extensively on the ability to estimate. Two aspects of this are important. The first is the ability to judge whether the result of a calculation which has been carried out or a measurement which has been taken seems to be reasonable. This enables mistakes to be detected or avoided; examples are the monthly account which is markedly different from its predecessors or the measured dose of medicine which appears unexpectedly large or small. The second is the ability to make subjective judgements about a variety of measures.[16]

The ability to judge whether an answer to a calculation is reasonable may be the most valuable mathematical ability that any student or adult person can develop, yet most students are missing it. I am not surprised, as mathematics learning worldwide focuses on precision and accuracy, and estimation and ish-ness are neglected. The committee in the UK shared that people in the workplace "rely extensively" on the ability to estimate. I have used my own ability to estimate the answer to number calculations and other mathematical problems many times over in my life.

Suzanne Downes, who teaches mathematics in international schools, shares this reflection:

I am starting to feel sad about young and older students that when asked to divide 272 by 8 mentally, they will try and do long division, without a real feel for whether the answer makes sense or not, inside their heads. Same for adding or subtracting mixed numbers. When asked to add 19¾ + 27⅓ many students change these into improper fractions with a common denominator. These students will lose most sense of the numbers on the way. Is it that we don't take enough time for teaching true understanding? Is there not enough time for instilling joy of number sense and logical thinking?

Suzanne is not alone in her concern for students producing non-sensical answers, lacking the important ability, highlighted by the UK committee, to think what the answer should be—the ish number, helping them know if their answer is reasonable. This is an issue that affects students in every grade and every level of mathematics. Suzanne poses it as a dilemma and asks if it is due to taking insufficient time. That may be part of the

problem, but I have a solution that helps even this persistent issue, that any of us, teachers or parents, can use at any time. This idea is incredibly impactful and can be used with any learner. Even adults in the workplace will be helped by this suggestion.

Bring Ish Numbers and Shapes into Math Teaching, and Value Them!

My proposal is that before anybody works on a calculation or any other mathematics problem, they think about what the answer will be—they come up with an ish answer. For example, Suzanne shares the example of 272 divided by 8. If I needed the answer to this in my life, I would think that $8 \times 30 = 240$, so I would estimate a little over 30. My ish number might be 32. I can only work out this ish number because I have developed number sense—an approach to number that many teachers will lament their students lacking. If you ask students to make an estimation before they calculate, you will probably find that students work out the exact answer and then round it, to make it seem like an estimate. They go to lengths like this because they have not been given enough opportunity to estimate and to create ish numbers. If people stopped to think about the approximate answer before they calculated, they would be helped in their development of number sense, they would avoid giving non-sensical answers, and they would learn to appreciate ishness, a critical tool for living.

Asking people to consider an ish answer before they work on mathematics can be used with all sorts of problems. For example, you might ask a student who has been asked to plot $f(x+3)$ on a graph, *What do you expect it to look like?* When we ask students to think about what they expect, before they work on any mathematics, we will be doing something that is extremely valuable for them. When I was writing this book I made a beautiful mistake. I thought I sent an email to a cognitive psychologist friend of mine, to ask him his thoughts on the cognitive processes involved when ish-ing numbers. But I accidentally sent the email to someone with a very similar name—a neuroscientist at Stanford. He kindly wrote back sharing that he was probably not the intended recipient of my email but he also shared how fascinated he was with my question, about the process of ish-ing numbers, and that the two processes—of ish-ing or being precise with numbers—probably involve different brain areas: the frontoparietal control network (FPCN) and the default mode network (DMN). Neuroscientists are fascinated by the question of brain connectivity and communication—knowing that enhanced functioning comes from different areas of the brain working in synchrony.[17]

This act of predicting what you will calculate or see sounds like a small classroom or home action, but it does something very important—it causes the brain to pull back from the detailed focus that it is in, and move into a different mode. This is a move from precision-focused thinking to a "big picture" mode or from micro to macro thinking.[18] Mathematics learning would be greatly improved if students learned to use these two different modes of thinking. When students are estimating, working with ish numbers, they are in a big picture mode, an important place to be. They can also work with precision, but it is the interplay between the two modes—precise and ish numbers, or focused and big picture modes—that is so valuable.

Precision is not unimportant, but ish numbers—and shapes—are equally important, yet they are almost completely neglected in teaching students mathematics. These unassuming numbers and shapes could help students immeasurably, giving a route to the development of number sense and shape sense. They can also help soften the sharp edges of mathematics, which is all many students need to jump in and engage. Teachers can use ish when talking with students who are nearly there, but not quite there. When we bring ish into our lives we are more protected from the dangers of perfectionism—a damaging mindset, and from binary thinking. Some people have asked me how asking students to "ish" numbers is different from asking them to estimate. It is a language change, but it is an important one, and this is why. When we ask students to estimate they think they are being asked to perform another mathematical method. But when we ask them to ish numbers they feel free—and they are more willing to share their ideas—at the same time as developing number sense. Go to Mathish.org to see how an ish perspective improves student engagement and understanding. We probably all need a little more ish in our lives.

The need for students to estimate numbers also emerges from a fascinating area of neuroscience concerning a part of the brain called the "approximate number system" or the ANS. Psychologist Darko Odic and neuroscientist Ariel Starr argue that before children even start school, they have an "intuitive, abstract and flexible" sense of number that draws from this area of the brain.[19] It is the first means by which people understand numbers, and researchers point out that the ANS exists across cultures, ages, and even species of animals. As its name suggests, this part of the brain focuses specifically on approximations of numbers and—notably—it predicts students' achievement in mathematics for many years ahead.[20] Yet we do little to cultivate this brain area and important ability, especially if schools do not value ish-ness and only focus on precision. In the next chapter, I will share a lovely activity, easy to conduct with any learners, anywhere, that helps develop the approximate number system.

Our journey together thinking about the diversity in numbers and the need to play with them and embrace their ish-ness starts with the nature of numbers themselves. If people spent time seeing numbers, exploring number patterns and ish numbers, and playing with numbers flexibly, they would develop great mathematical power, and young children would get the best mathematical start possible. But this is not just for young students; adults of any age can develop a different relationship with mathematics when they start to see numbers in a playful way.

Later we will see more ish shapes. If we want mathematics to be a useful lens with which students view the world, it is really important that we value the ish nature of numbers and shapes, celebrate them, and build upon them, to a greater extent in classrooms. When I first shared the ideas of ish at a mathematics conference, an audience member recommended that I read Peter Reynolds's children's book *Ish*. In this beautiful book, which has inspired so many teachers and students, the author shares the story of Ramon, a boy who renews his self-belief and motivation to keep drawing when he thinks "ishly." Reynolds captures in his book the essence of something I hope to convey with math-ish. When we allow students to think "ishly" about math, it frees their thinking and invites them into a new domain of mathematical creativity and diversity.

Most students do not experience mathematics in ish, playful, visual, or diverse ways, to their detriment. That is bad enough, but things really start to go awry when people learn operations—adding, subtracting, multiplying, and dividing—and then fractions. These are important and weighty topics that will be needed throughout people's lives, and we will explore them in creative ways soon. Before that, we will consider an area of mathematics that has recently exploded onto the scene and plays an important role in preparing our young people for their futures, creating another opportunity to see the beauty of mathematical diversity and ish-ness. College professors have named this one of their top three content areas in mathematics: data literacy.

Data Literacy

About ten years ago, the world shifted in a significant way as larger and larger amounts of data started to be collected and stored. By 2020, there were ten times as many bits of data in the world than stars in the universe. Data is now used in all businesses, big and small; in sports analysis, health care, education, and entertainment; and in just about every other field you can think of. The US Bureau of Labor Statistics considers data science and analytics one of the top twenty fastest growing occupations, and they estimate an increase in demand of more than 30 percent over the next ten years.[21] There is no doubt that today's children will be entering a data-filled world when they leave school, and almost all employees will be more effective if they can read and interpret data.[22]

But helping students to accurately interpret data and data visuals will not just help them with their employment. As soon as young people start accessing social media and the internet, they become vulnerable to the circulation of misinformation. We need to help them learn how to separate fact from fiction, to make sense of the different data visuals and information they are sent, including those that are intended to mislead them. I regard this as an important equity imperative. If we do not help our young people make sense of data and data representations, we leave them vulnerable to the post-factual world that is lying in wait for them. Even our youngest learners can and should be introduced to data, and in the remainder of this chapter I will share some of the big, critical ideas in data literacy, important for everyone in the world, and some ideas for encouraging data literacy in families and classrooms.

When Steve Levitt asked me to present on his Freakonomics show, he told me about a group of leaders working to help bring greater data emphasis into K–12 education. The group included Levitt himself; Arne Duncan, former Secretary of Education; Nate Silver, statistician and creator of FiveThirtyEight, who provides insightful election analyses and more; and Eric Schmidt, former CEO of Google. I agreed to help the group and started to work with my team at Stanford to create data resources to support teachers. In highlighting the data understanding that students need, I was aware that teachers had no space to add any new content to their schedules, so we helped create something more useful than content—a data awareness that would enable teachers to infuse data ideas into the content they already teach.[23] Children as early as kindergarten can start to develop data literacy that will help them read the world. As students move into middle school, they need to be making sense of data and probability as part of their mathematics standards. In high school, students can take introductory data science courses and move to a pathway that is focused on data science and statistics as an alternative to algebra, trigonometry, and calculus.[24] Fortunately, universities are also shifting to value these different pathways as highly as the more established calculus pathway.

Systemic Inequities: A Math Reality

Harvard University, which, like many other schools, supports the broadening of mathematics courses, communicates that high school mathematics courses should focus on conceptual thinking and encourage students to use reasoning to critically examine the world.[25] A course in data science is an ideal opportunity for students to learn these important skills.

It may seem a small step to allow students to take a high school pathway that is focused on data science and statistics, but this is the first time the content of high school mathematics has changed since the 1800s. In 1892, a group of ten white men (called the Committee of Ten) developed the mathematics curriculum that students should learn in schools.[26] It is baffling that we still teach this same mathematics, even though the mathematical needs of the world have changed dramatically. Of course, the change is being greeted with significant pushback from traditionalists who believe the only pathway that should be offered to students is one of algebra and calculus.

I like and appreciate calculus. It is a powerful set of ideas that, as Steve Strogatz points out, has enabled us to have cell phones, computers, microwave ovens, radio, and television.[27] I took calculus in high school with a wonderful mathematics teacher who invited her students to discuss ideas. She was the first math teacher that gave me this opportunity, and it changed everything for me. But there is a big problem with the calculus pathway in the US: there are more courses in front of calculus in high school than there are years of high school. For students to enter high school on a pathway that leads to calculus, they need to take algebra in middle school. This has led school districts to set up two middle school tracks, one that leads to algebra in eighth

grade and one that does not. Middle schools often start these different pathways in sixth grade, using test data from fourth grade. This means that a narrow test taken by children aged ten or younger decides what course they take in twelfth grade and, from there, what college or other future is open to them. This sorting of young students has led to indefensible racial and social inequities. Forty-six percent of Asian students take calculus, compared with 9 percent of Latine students and 6 percent of Black students; overall, only 16 percent of students take calculus in the US.[28] This is because most students, particularly students of color, are pushed out of the high-level pathway from a young age. Only half of the high schools in the US (typically, those in more wealthy areas) even offer AP calculus, making it problematic when colleges use that course as an entry requirement.[29] Even the 16 percent of students who take calculus in high school are not well served. David M. Bressoud, mathematics professor at Macalester College and former president of the Mathematical Association of America, examined a large dataset of over 800,000 students and found that more than two-thirds of the students taking calculus in school retook it or took a lower-level course in college.[30]

The solution to these problems of systemic racism and low participation is not, of course, to eliminate calculus; it is to rethink the sequence and content of the courses. If high schools did not have four courses in front of calculus (algebra—geometry—algebra 2—precalculus), then middle schools would not need to set up different pathways that draw from elementary school achievement. I was one of five writers of the 2023 California Mathematics Framework, a set of policy guidelines that was unanimously voted into policy by the California State Board of Education. One of the recommendations of the framework is to rethink and streamline the content in high school courses.[31] If the content that is no longer needed in the modern world were removed from courses, and if the remaining content were taught, as Harvard recommends, so that students could "use mathematical reasoning to critically examine the world," we would have many more mathematically empowered students. The Mathematical Association of America recommends that schools stop what they call "the rush to calculus" and leave it to be taught in college.[32]

Another way to disrupt these systemic inequities is to offer data science as a high school course. Students could take data science as a third-year high school course, without being advanced in middle school. Whereas the majority of students who were put on a lower-level track were previously on a path to mathematical nowhere, they can now be on a path to a beautiful mathematical somewhere. An ideal course to follow a course in data science is statistics. In the year after Steve Levitt asked me for help in bringing data science into K–12 education, my team at Stanford developed a high school course in data science, with the advice of a team of data scientists and professors of data and statistics. The course uses only free tools.[33] As I write this, there are five data science courses available to students in the US. In the second year of the course, we already had over 160,000 students taking it. Their teachers shared that the students were

- 46 percent girls and nonbinary people,
- 57 percent students of color, and
- 68 percent students who had not been mathematically accelerated.

The free course provides an important mathematical option for this diverse group of students, whatever their mathematical past. Research on students taking data science shows that they take more mathematics and are more enthusiastic about STEM and higher education at the end of the course.[34]

Sadly, traditionalists campaign against this opening of pathways and broadening opportunities for people who have not typically had a chance to study and work in STEM. The battle to open mathematics to more people has raged throughout history.[35] The opposition comes from those who have experienced great success in mathematics. A cynic might say that the idea that anyone can learn mathematics to high levels is threatening to the identities of those "special" people who have shown that they are superior. Another cynic might say that they like a system of mathematics education that segregates children by race, gender, and social class. These disturbing ideas are certainly not held by all high-level mathematicians; it is my conjecture that those who campaign against the opening of mathematics for reasons of racism and bias are a small but loud group—which I will return to in chapter 8.

Data science in K–12 classrooms is not only a new possibility for high school students but also an exciting new lens with which to view mathematics, offering all teachers the chance to diversify content for students of all ages. When teachers take a data perspective, they can teach the same numbers they have always taught students, but now the numbers have meaning—they represent something real in the world. Of course all of the numbers students meet in the world are ish numbers, and that is something to be acknowledged and celebrated. For example, you may want to teach people about decimal numbers. Instead of giving students a worksheet of numbers, you could ask them to measure objects in the room, or around the school, and record the data they collect in a data table. Or perhaps they could measure a plant's growth. The world is filled with naturally occurring decimals and ish numbers!

Get Curious About Data

Data does not have to be a number; some data is "categorical" or "qualitative." If you decided to record people's favorite colors, you would be collecting categorical data.

Another important way in which numerical data varies is whether it is continuous or discrete. Continuous data occurs when the numbers make sense in between the main values. For example, if you collected data on people's heights, you could draw a line between a height of 1.5 meters and a height of 1.8 meters, because there are meaningful numbers between 1.5 and 1.8. An example of discrete data is the number of legs in your family, including your pets, or the number of siblings a group of people has. You cannot have 2.5 siblings or legs.

These forms of data are all around us, and students in classrooms can have a lot of fun exploring the data in their worlds. Doing this not only gives the ish numbers they will encounter meaning; it gives mathematics meaning. As students become comfortable with data, they can start conducting data investigations, which start with a question they have asked, as represented in figure 4.12. A feature of data investigations that I particularly like is that they invite people to be pattern seekers, an approach that helps with all mathematics. When students investigate and find patterns, they are invited to develop meaning from their work and communicate their findings. This process is beautifully cross-curricular and can satisfy standards in math, science (or whatever is the topic of the investigation), English, and more.

An ideal start to a lesson or a family conversation is what we call a "data talk." Students are invited to "notice and wonder about" a data representation. Sharing these data visuals with young people and helping them make sense of them is critical in helping them develop the data literacy that will protect them

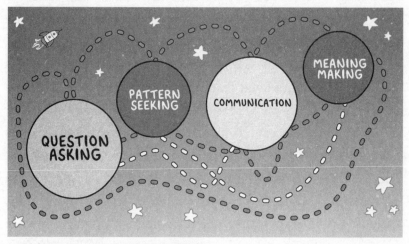

4.12 Investigative data science process

from the post-factual world that is lying in wait, ready to mislead them.[36] Data representations can be found everywhere: magazines, newspapers, social media, and websites. When I run data talks in classrooms, I encourage students to pay close attention to the data and the source of the data.

Beyond illustrating the power of thinking about data, data talks show students of all ages the creativity that is possible in modern-day data visualizations. One example of the traditional curriculum's being out of step with the needs of society is that schools ask students to consider line graphs every year for five years. In the world today, data visuals are extremely creative, showing aspects of data that line graphs cannot show. One of my favorite data talks shares data on the positioning of basketball legend Stephen Curry when he takes shots from different places on the court, shown in figure 4.13. The same information could be plotted on a line graph, but the representation on a basketball court shows much more.[37]

When we first shared this data visual on youcubed, I realized that we could also share one from my favorite sport—soccer. Data is now of huge value to sports programs, who use it to select players and to improve the performance of players and teams. When looking for data visuals highlighting soccer, I discovered Michael Poma, who was the data analyst for the women's soccer program at James Madison University and is now the data analyst for the women's professional soccer team, the Houston Dash.[38] His role, to me, highlights the range of jobs available to people with data understanding. He shared with us the data representation in figure 4.14, which shows a soccer goal and players' shot placement when taking penalty kicks. (For those who do not follow soccer, a penalty kick is awarded when a player is fouled inside the box that surrounds the goal. Players shoot at

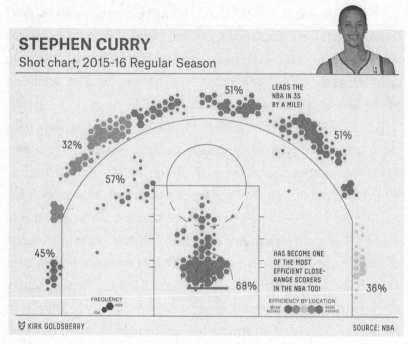

STEPHEN CURRY
Shot chart, 2015-16 Regular Season

51%

LEADS THE
NBA IN 3S
BY A MILE!

51%

32%

57%

45%

HAS BECOME ONE
OF THE MOST
EFFICIENT CLOSE-
RANGE SCORERS
IN THE NBA TOO!

68%

36%

FREQUENCY

HIGH
LOW

EFFICIENCY BY LOCATION

BELOW
AVERAGE

ABOVE
AVERAGE

KIRK GOLDSBERRY   SOURCE: NBA

4.13 Data representation of all Stephen Curry's shots in open play in the 2015–2016 season

the goal, with only the goalkeeper in their way.) PSxG, or post-shot expected goals, measures the probability of a goal after the ball has left a player's cleat. In the data visual, the density of the colors shows the success of the shots players take.

I have shared two data visuals that focus on sports—a topic filled with data and data analysis that can really engage students—but you can choose any topic for a data talk, such as deforestation, costs of housing, popular dogs, protection from viruses, and more.[39] Data talks share data in visual and creative ways that prompt rich conversations and, importantly, the development of data literacy as people learn to read the data.

Being from the UK, I have done my fair share of Atlantic hopping since I moved to the US. Given my transatlantic expe-

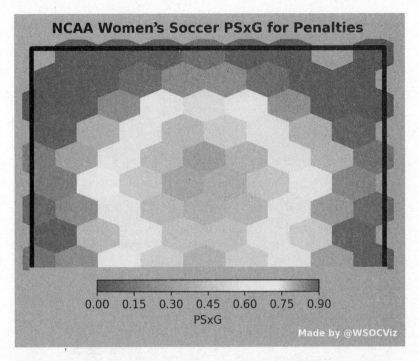

4.14 Data representation of penalty shots taken in National Collegiate Athletic Association (NCAA) Division I women's soccer games between 2017 and 2019 (approx. 6,500 games)

riences, I was particularly interested to learn about two award-winning designers, Giorgia Lupi and Stefanie Posavec, who live on opposite sides of the Atlantic. Stefanie lives in London, and Giorgia lives in New York. To stay connected with each other, the two designers sent each other a postcard every week for a year that shared aspects of their lives through data. They called this project "dear data," and they chose a range of topics such as smiles, laughter, or indecision.[40] They collected different pieces of information about their topics, which are called variables. More than two variables are known as multivariable data, a central idea in data literacy and data science. For example, one week they focused on goodbyes (fig. 4.15), and the variables they collected were the type of goodbye, who it was with, and where it was said. Their beautiful data visuals are now on websites and in books.[41]

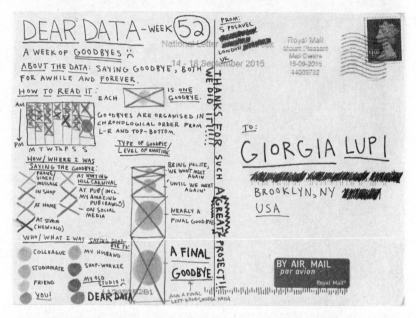

4.15 Week of goodbyes: postcard from Stefanie Posavec to Giorgia Lupi
dear-data.com

Inspired by the work of the two designers, I incorporated a similar activity into my teaching at Stanford and in the lessons on our website and in our high school course in data science. I invited students to collect data on anything in their lives, using three or more variables. In the classroom, they were given time to make data representations. This activity changed students' perspectives on data and mathematics in important ways.

Kira Conte was an undergraduate who took my course at Stanford; her mother is a youcubian who shared with her daughter the importance of mindset. Kira first used her growth mindset to get into Stanford and then started researching mindset and contributing to the field. When I gave my undergraduates the challenge of collecting data on their lives—collecting at least three variables—Kira chose to focus on the ways she interacted with her dog, Daisy. The data she collected included the type of interaction, the reaction of Daisy, and the time of day. The visual she created illustrates the creativity possible when working with data.

In the first unit of our free data science course, students are invited to collect data on their own lives and create data visuals with their peers.[42] Students who experience this activity have told us that it is the first time they have ever brought their lives into a mathematics class. One of the students decided to keep

4.16 Data representation showing Kira's interactions with her dog, Daisy

a record of their food and drink, which caused them to notice they needed to eat more healthy food! Another student recorded the activities of their hamster, using a phone recorder, over twenty-four hours. Others coded the music they listened to, the languages they spoke at home, and their use of swear words, which they illustrated in amusing ways! The students reported feeling connected, for the first time, to the mathematics they were learning because of its personal relevance. Any teacher can include this unit in their teaching, giving students a chance to consider data and learn mathematics with meaning.

The students in the youcubed course are introduced to data through collecting data on their own lives to open their minds and their perspectives on the ways data can be used. They go on from this opening activity to learn ways to interrogate large datasets, to construct and use mathematical models, and to become effective data analysts. A study of students taking the data science course shows that they found the course to be different from their previous experiences of mathematics—not in the level of rigor or the difficulty of the concepts—but in the ways they could access the concepts and in the diversity of ideas they encountered.[43] Their previous experience of mathematics was one of right and wrong answers, and of exactness which contrasted with their learning of data science, which is more typically about the use, application, and interpretation of a broad set of ideas and the interpretation of ish numbers. These are all important components of mathematical diversity.

Mathematical diversity gives students deeper mathematical understandings, even of traditional concepts such as algebra. This was shown in an assessment taken by students who were taking a course in algebra 2 and/or a course in data science at multiple high schools across three school districts. The students in our study were asked to consider the relationship between different variables—daily calories, life expectancy, and infant mortality—and to set up a linear model examining the relationships. This content was chosen because it is central to algebra courses as well as data science courses. The students in data science courses scored at significantly higher levels than those taking algebra 2 courses ($p < 0.001$). The P-value shows that the achievement difference between the two groups being due to chance is less than one in a thousand. Although the students in algebra 2 courses should have been able to investigate the variables and create linear models, they seemed unable to work with real data.

Recall from the beginning of the chapter that we asked college professors to identify areas of mathematics that are important for incoming college students. One of the top three they highlighted is the area described in the above assessment: linear equations. This important mathematical concept—showing how variables relate to each other—is taught inside data science and algebra courses, with applications for people living their lives beyond the classroom.

Linear Relationships

During the COVID-19 pandemic, our news screens were filled with linear models showing how the virus was spreading, the danger it posed, and the ways we could minimize its spread with vaccines. Some people were able to use their data literacy to understand the information presented; many others struggled to do so and were misled by misinformation that left them ill-equipped to protect themselves or loved ones. Much important information is presented as linear relationships, as you have likely seen in data about mortgage rates, health and fitness, sports, weather, and more.

As we consider linear relationships in the world, one important concept taught in data science courses is worth knowing about: the difference between correlation and causation. The graph in figure 4.17 shows the num-ber of shark attacks on one axis and the sales of ice cream on the other.

4.17 Ice cream sales and shark attacks

Many people would look at this graph and see that the two variables are correlated—thinking that a change in one *causes* a change in the other, which would mean the relationship is *causal*. But in fact the apparent relationship is caused by a third variable—known as a confound-ing variable. You may be able to work out what that confounding variable is.

The confounding variable is the hours of sunshine. The hours of sunshine cause shark attacks to go up because people flock to beaches and swim in the ocean; the same hours of sunshine also cause ice cream sales to go up. This makes the two variables—shark attacks and ice cream—correlate, but they are both caused by something different, the hours of sunshine.

So much data appears to be causal, but, in fact, their similarity is caused by one or more confounding variables, which has led Tyler Vigen to develop a hilarious website sharing what he calls "spurious correlations."[44] One of the correlations he shares is between the consumption of mozzarella cheese and the number of doctorates awarded in civil engineering.

When I share the data graph in figure 4.18 with audiences and ask them what they see, they speculate that the engineers probably ate a lot of pizza when studying for their doctorates! The website does not share what the confounding variable is; they probably want us to keep guessing!

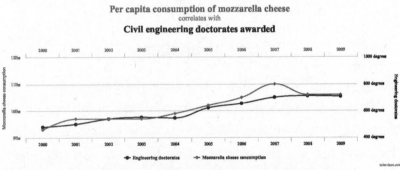

4.18 Consumption of mozzarella cheese and engineering doctorates
US Department of Agriculture and National Science Foundation; tylervigen.com

In later chapters, we will look at more examples of linear relationships, as they are important to our lives. In this chapter my examples have all included data because I am committed to the idea that all mathematical topics, including algebra, can be enhanced with data taken from the world. In the years that I have been sharing data with educators, I have found that when people invite data into their lives and their teaching, they start to see data and mathematics as more ish-like, and, with that, more cool, accessible, and pervasive.

DATA AWARENESS

As the investigation of data has become more widespread in the world, so has a disturbing phenomenon—the number of people working to spread misinformation, some examples of which have had major consequences.[45] To protect yourself, and others, from the spread of misinformation, I recommend asking yourself these important questions whenever you see data or data visuals:

- Who produced the data? What is their goal in producing it?
- Is all the data being displayed? If not, what is omitted?
- Are the axes/legend sensible? Or are they built to emphasize a point?
- What relationships are evident?
- Are the relationships causal or just correlated?

If you consider these ideas as you go through your life, you will be protected from villains who seek to mislead you (including banks and moneylenders), and you will become data empowered!

In this chapter we have considered mathematics in the world, including the three mathematical ideas college professors say are the most important—number sense, data sense, and linear relationships—and what they look like when they are mathematically diverse. In the next chapter, we will expand on these ideas by looking through a different lens, which is an important part of mathematical diversity. I am passionate about sharing this lens—the importance of seeing mathematics visually, which leads to incredible beauty and creativity.

MATHEMATICS AS A VISUAL EXPERIENCE

It was 2016, a year after we taught the first youcubed summer camp to a group of middle school girls and boys on the Stanford campus. We knew that the students scored at significantly higher mathematics levels at the end of our camp, but we wanted to know if there had been any long-term impact.[1] We caught up with our former students by visiting them in their schools, and we interviewed them in pairs. The students talked in different ways about how much they had been helped and how knowing that mathematics was a subject that could be approached in diverse ways had changed their learning going forward. One reflection particularly stood out for me. It related to an activity we had given the students called painted cube.[2] The students were shown an image of a 4 x 4 x 4 cm cube, made up of smaller 1 cm cubes, dipped into a can of blue paint. We asked how many of the smaller cubes would have 0, 1, 2, 3, 4, 5, or 6 sides that were blue.

This is an activity that leads to high-level algebraic reasoning. The students were given

1 cm sugar cubes and built the 4 x 4 x 4 cubes in their groups. The physical building of the larger cube meant that students were activating different brain pathways, developing the areas of their brains responsible for numerical, visual, and physical processing.[3] A year later, I talked to one student, Jed, to ask him how the camp had impacted him, and his reply surprised me. Jed told me that he was now in geometry class, and he had been helped all through the class by remembering the 1 cm sugar cube he had held in his hands in our camp. He said that he could remember what it looked like, as well as the physical feeling of holding the cube in his hand, and it had given him a picture in his mind of what "1 cm cubed" meant. He said that he was using that memory to solve many of his geometry questions. One example he shared was a math question asking students to estimate the volume of their shoes. Jed said that he imagined his shoe filled with the 1 cm sugar cubes. I now know that Jed was describing the kind of "mental representation" that Anders Ericsson and Robert Pool describe as critical to the development of expertise.[4]

Mental representations don't have to be physical objects, but they do require more than most students ever receive in their

mathematics classes. Over the next few chapters I will describe and share several mental representations that can help students immeasurably. Some of these, I think, will surprise you.

Anders Ericsson, a world expert on expertise, looked at the nature of learning and performance in those achieving at the highest levels, in many different fields, including chess, sports, and school. He and Robert Pool, his coauthor, are probably most famous for describing an important learning condition as "deliberate practice."[5] The most important quality of deliberate practice is the opportunity it provides to develop mental representations. A second important quality of deliberate practice is the opportunity it gives students to struggle, as Ericsson and Pool describe:

> You don't build mental representations by thinking about something; you build them by trying to do something, failing, revising, and trying again, over and over.

MENTAL REPRESENTATIONS

Ericsson and Pool describe mental representations with examples of American football and actual football (otherwise known as soccer!). In both sports a novice may look at the field of play and see chaos in twenty-two players scattered all over the pitch (fig. 5.1).

But an expert in soccer will see patterns, and those patterns will help them understand the way the game works and the progression of important moves, as figure 5.2 illustrates.

Ericsson and Pool point out that to novices, soccer appears to be a swirling chaos moving toward the ball, but to an expert "this chaos is no chaos at all, it is a beautifully nuanced and constantly

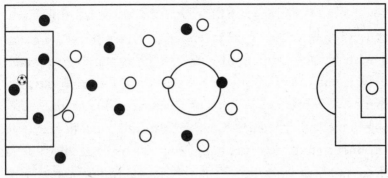

5.1 Soccer pitch with 22 players

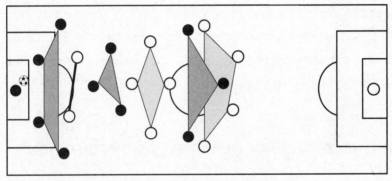

5.2 Soccer pitch with 22 players, with mental representations
of their patterns of action

shifting pattern."[6] Experts have a highly developed ability to interpret the pattern of action on the field.

I have noticed this in my own understanding of each sport. I
grew up watching football (soccer!) in England and from the age
of four sat on the terraces of West Bromwich Albion's ground—
the Hawthorns. (In the 1970s I was extremely proud that my team
became the first in history to have three Black players. My club
has continued to lead the way in promoting the value of racial and
cultural diversity in sport.)

When I moved to the US, I spent many years ignoring American football, mildly amused that a sport involving holding,

throwing, and catching a ball in your hands is called football. But over recent years, I have taken an interest in the game, starting with watching Stanford games, then expanding to watch the 49ers and eventually other teams. When I started watching American football, the position of the players did not mean anything to me, and it seemed like there was too much to take note of to understand what was happening—it really seemed like a chaotic jumble of players. As I have learned more and started enjoying the game more, I have not learned facts or procedures about American football, but I have learned to see patterns, and those patterns are now mental representations in my mind. Those mental representations allow me to see and understand what is happening in the game.

THE NEUROSCIENCE OF MENTAL REPRESENTATIONS

Ericsson and Pool are not the only researchers who talk about the importance of mental representations or models. Cognitive science has a long history of establishing the value of mental models for learning and performance. Interestingly, this is now emerging as a key finding from neuroscience: our brains function by producing models of our world. Neuroscientist Jeff Hawkins has spent his career studying the role of the brain's frontal cortex.[7] This important part of the brain is responsible for cognitive skills, problem-solving, higher-level functioning, and social interaction—a wide-ranging, critical set of skills. Hawkins has found that our brains function by creating models of the world—which are constantly adapting as we move from one experience to the next. This connects to the cognitive science showing the value of mental models for our learning and understanding. It is no wonder that students need mental models to ground their thinking and knowing, as this is how the brain functions not only in learning but in life.

I started this chapter with the example of my student Jed sharing the value of seeing and holding a 1 cm cube of sugar. Jed's mental model was physical, and there is considerable evidence that students interacting with physical representations of mathematics is extremely generative for their learning.[8] But students and teachers can also create visual representations, which are another important form of a mental model. As a thought experiment, think for yourself about the physical or visual representations you were encouraged to develop as mental models for mathematical ideas. If you have had a typical math experience of numbers and procedures, you may not be able to think of any. That is because, for most people, math is an almost entirely symbolic, numerical experience. When visuals do appear, they are often sterile pictures drawn by publishers, showing bisecting angles or circles divided into slices, which do not help students

develop their own visual or physical models of mathematical concepts. If you are an adult who had few or no opportunities to develop mental models of mathematical ideas, I hope that the rest of this chapter will be particularly helpful for you.

When researchers compare the brains of mathematicians with similarly high-achieving academics who work in non-mathematical fields, they find something fascinating. We might assume that when people are thinking numerically, they are thinking "with language" and that high-level mathematical reasoning would draw from language-processing areas of the brain. What the researchers found was that the brain activity that separates mathematicians from other academics comes from visual areas of the brain—and this was true whatever the mathematical content.[9] Not only geometry and topology but also algebra and other calculations caused activity in the visual brain areas, with minimal use of any language areas. This led the researchers to offer the possible explanation that the mathematicians' achievement came from early childhood experiences with "number and shape"—which, I would add, probably helped them develop mental representations of mathematical ideas from a young age.

The groundbreaking work of neuroscientist Vinod Menon has shown that when people work on a mathematics problem, even an abstract number calculation, they can be helped by five different brain pathways (fig. 5.3).[10]

When *National Geographic* magazine summarized an investigation into the nature of a "genius," they added some useful information on these different pathways. They considered highly accomplished people they described as "trailblazers." These people, who stood out for their "meteoric contributions" to their

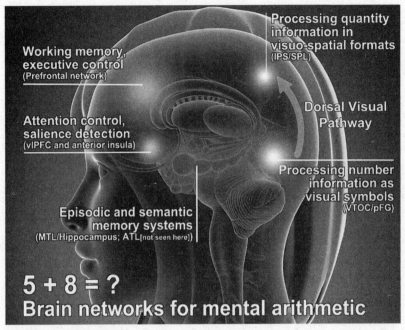

5.3 Brain areas available for mathematical thinking

fields, included scientists Albert Einstein and Marie Curie, comedian Anne Libera, and mathematician Terence Tao. A conclusion of this fascinating discussion is that the accomplishments people ascribe to being "genius" actually come from a complex combination of circumstances that include learning opportunities as well as culture, geography, privilege, and, of course, brain development. Notably, what separates the brains of trailblazers from regular people is the amount of connectivity they have between brain pathways and the greater development of visual areas of the brain.[11] Two of the brain pathways, which focus on visuals, are at the back of our heads. When we encounter a problem in numbers and see a visual representation, or a well-worded description, connections are made between brain regions. Researchers have shown that students achieve more highly when they are given mathematical work that involves both numbers and visuals.[12]

Fortunately, those of us who teach mathematics have many opportunities to introduce ideas to students in ways that stimulate different brain pathways and connectivity—for example, when we invite students to see concepts not only as numbers but also as words, visuals, physical representations, tables, algorithms, models, and movement. We should aim to give our students, our children, and ourselves a connected, multidimensional experience of mathematics that allows different brain pathways to communicate and connect.

GROUPITIZING

The fields of mathematics education, cognitive science, and neuroscience have for decades contributed studies showing that student learning is enhanced by visual and physical representations of mathematics.[13] Despite these different and extensive bodies of work, mathematics classrooms everywhere and mathematics textbooks published by the most powerful and dominant companies are filled with pages of numbers. Traditional mathematics instruction will be slow to change, but instructors can start now to help their students develop powerful mental models; models that are visual and physical and foundational for a rich and deep understanding of mathematics. When students learn to create and use visual and physical mental models, they are invited into the realm of mathematical diversity, to see mathematics in different ways and experience mathematics as a diverse set of ideas. The examples I will share come from the areas of number,

multiplication and division, fractions, and algebra—spanning the K–12 grades. These examples are intended to help teachers, parents, and any reader see that they can—and should—create opportunities for learners (including themselves) to develop their own mental models in any area of mathematics.

Seeing Numbers

Neuroscientist Bruce McCandliss, a colleague of mine at Stanford, focuses his work on education and learning.[14] He and his team have produced some stunning research evidence showing that the way young students see and group numbers predicts their achievement in state tests for years ahead and even ameliorates the effect of low household income.[15] Prior to their work, researchers and educators had noted the importance of what is known as subitizing. Humans have a natural ability to look at a group of dots or other objects up to four and say how many there are without counting them. This ability usually shows up during kindergarten and develops throughout the later grades. McCandliss and his colleagues introduced the term *groupitizing* to describe the ability to group a larger collection of dots using subitizing. For example, if someone saw the visual in figure 5.4 and was asked how many dots there are, they may say there are 10, as they see (subitize) groups of 4, 4, and 2.

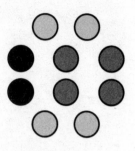

5.4 Groupitizing: 4 + 4 + 2 = 10

In their study of 1,209 K–8 students, McCandliss and his colleagues showed that the extent to which students could groupitize predicted their achievement in state math tests above and beyond their "fluency" in numbers or their achievement in arithmetic. The results were consistent all the way through grade 8. Figure 5.5 shows the important relationships the researchers noted.

The researchers found that household income (data they collected as a measure of equity) has an unfortunate relationship with state test scores, showing a standard regression coefficient of 0.67 (the bottom part of the diagram), even after controlling for mathematics achievement. But the model also shows that groupitizing is almost as impactful, with a standard regression coefficient of 0.56. The bottom row of the diagram shows that when students have learned to groupitize, the impact of low household income drops to 0.25. This data is so important because it shows

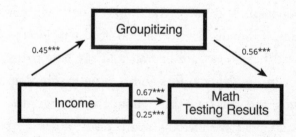

5.5 McCandliss et al.'s visual representation of the relationship between groupitizing, household income, and state math test results

that groupitizing significantly boosts students' mathematics achievement and reduces the impact of systemic inequities—two very good reasons why the practice can and should be taught to students.

5.6 Seven dots

One way to teach groupitizing is an activity called a dot card number talk. In this activity teachers or parents show a collection of dots and ask learners how many they see. It is important to tell learners that they have only a few seconds to look at the image, so they will need to group the dots rather than count them. I usually tell my classes that our brains naturally want to group the dots, so it is something everyone can do.

When I ran one of these activities recently, with a room filled with middle school girls, the students came up with twenty-four different ways of grouping seven dots (figs. 5.6 and 5.7).

5.7 The 24 ways middle school girls grouped the seven dots

I always represent the different groupings and give each method the name of the student who offered it, as shown in figure 5.8. I also ask students to share the number sentence that goes with their visual, allowing us to see the many different ways numbers can be formed.

The research by Bruce McCandliss and his team is stunning. We spend so much time in elementary education teaching students numbers and operations, when giving them experience to

5.8 Our first youcubed camp, where I shared a dot card number talk, showing the students' visuals with their names and number sentences

see the ways dots can be grouped into different numbers improves their mathematics achievement, even on narrow state tests, to a much greater degree. As students learn to groupitize dots, they develop mental models that they can return to every time they think about numbers. This single study should, in my view, cause a rethink of the mathematics priorities in elementary schools nation- and worldwide. In other research, Bruce and colleagues show that mathematical activities that are built around Cuisenaire rods, which I shared in chapter 4, bring about important changes in students' understanding of numbers, fractions, and foundational ideas such as equivalence. Importantly, the rods give students mental representations of these abstract ideas, which extend through to their learning of algebra.[16]

Another important model for numbers is our fingers. Recent studies have produced incredible evidence showing the importance of finger perception to our understanding of mathematics. *Finger perception* is a term that neuroscientists use to describe the extent to which you know your own fingers. A test for finger perception is to put your hand under a book or table, so you cannot see it, and have someone lightly touch your different fingers. If you can identify all your fingers, you have developed finger perception. Researchers have shown that finger perception is a better predictor of mathematics achievement in second grade than test scores.[17] They have also stated that discouraging students from using fingers is akin to halting their mathematical development.[18] I am convinced that fingers carry immense value because they provide a physical model of a number line. Evidence for this claim comes from other studies showing that students who learn with number lines boost their achievement significantly. In one study, researchers noted that students arrived at the first year of school with differences in their number sense that related to household income, with students from less privileged homes having weaker number sense. This difference was completely eliminated when students played games with number lines for four fifteen-minute sessions.[19]

A number line shows a continuous representation of numbers, but it can be complicated for young learners because their fingers often land between the numbers, and they do not know what that means. A number path, as shown in figure 5.9, is a better tool for beginning students to develop a mental representation of numbers.

5.9 Number path

A classroom I visited recently had a giant number path wrapping around the walls of the classroom—students frequently looked up to it and used it when working with numbers.

When students are ready, they can use their fingers to develop a mental model of numbers. Students using their fingers to consider numbers are developing a model that they can carry around with them throughout their lives.

We are just starting to realize how important visual and physical models are for students' understanding of mathematics, and the emerging research from neuroscience and education should be given the highest priority.

MATHEMATICALLY DIVERSE OPERATIONS

As students are learning about numbers, it is important that they experience them not only visually and physically but also playfully. The opposite of this is to experience numbers and operations as a set of remote rules that must be followed. I would like to illustrate the difference with two classroom activities focused on the addition of numbers up to 20—an area of mathematics taught in the US in first grade. One of the activities exemplifies narrow mathematics; the other exemplifies mathematical diversity and the development of opportunities that allow students to create mental models.

Adding Numbers

Many textbooks used across the US and throughout the world present mathematics as a series of questions, as shown in figure 5.10.

Figure 5.11 is a different presentation of the same content. Students are introduced to a set of animals, all of whom have different numbers of feet, and they are invited to make "foot parades" with different numbers. For example, what animals would you include if you are making a foot parade of 20?

A teacher who was using the narrow worksheet in her classroom asked me a pointed question: *How can I have students in the same class if some can add to twenty and some cannot?* This is a reasonable question in the world of narrow

1 + 17 = ◯		4 + 12 = ◯
7 + 12 = ◯		9 + 10 = ◯
8 + 7 = ◯		2 + 9 = ◯
12 + 5 = ◯		10 + 2 = ◯
1 + 12 = ◯		18 + 2 = ◯
6 + 4 = ◯		9 + 8 = ◯
14 + 4 = ◯		2 + 17 = ◯
9 + 3 = ◯		19 + 1 = ◯
1 + 19 = ◯		13 + 1 = ◯

5.10 Textbook approach to number addition

math. Imagine first-grade students sitting at their desks faced with the worksheet—some, who have been tutored, may race through it without any thought. Others would be completely at sea, staring at the worksheet with panic and fear rising in their young minds. It is understandable that teachers don't want to give this content to students with different levels of achievement and present capabilities.

5.11 Animals to make a foot parade
J. Boaler et al., Mindset Mathematics, Grade 1

But if we break out of the world of narrow math into the world of mathematical diversity, creating opportunities for students to develop mental models, everything changes. The foot parade version of adding to 20 differs from the worksheet in at least three ways. One is the visual nature of a foot parade, which is significant not just because it is more visually appealing but also because it gives students something to count and a model they can develop in their minds. This means that students who want to count the legs can do so and others who can form the number bonds without counting can also do so. Another difference is that the students are interacting with something from the world that is interesting and engaging. In my experience, young children love to choose the different animals and display them on posters (fig. 5.12).

5.12 Foot parade poster

The third difference is that there are various ways to make each number, enabling students to create their own foot parade that they can feel proud of rather than racing to get the same answer as each other. With activities such as these, it does not matter if students have different knowledge—the breadth of the question and the different access points mean that all students can engage and learn. Additionally, the anxiety and boredom provoked by the worksheet are replaced by engagement and enjoyment, even though the exact same mathematics is being taught. This

is because we have broken out of the world of narrow math and entered the wonderful world of mathematical diversity.

Now all students—of different backgrounds and prior achievement—can engage with the task. Some of them may add the legs of the animals; others may use their prior knowledge of number bonds. Some may take the task to high levels, investigating how many different versions of 18 legs there are. We do not need to separate young students into different groups and classes, giving them damaging ideas about their own potential, if the mathematics tasks are diverse. Critically, the diversity in the tasks allows students to develop visual mental representations from which other learning can flow.

Multiplication and Division

As students progress through the years, they learn about multiplication and division, topics that are usually presented numerically. In the narrow version of a multiplication or division question, there is one valued method and one valued answer. But the diverse version of mathematics enables students to think

about the different ways they can multiply or divide and how to
see these ways visually, which creates important opportunities
for brain connections and mental models.

I do not think I need to share another example of a narrow
worksheet, as I did for addition. You probably have plenty of
multiplication and division worksheets burned into your minds
already. Instead, let's consider opening the multiplication question by asking you to consider 38 × 5 in your mind, without writing it down. When I first do this activity with students, they do
not know what visualizations to draw; they often have no visual
ideas at all. But over time they learn to visualize numbers, and I
share their methods and visual representations (fig. 5.13), which
they frequently tell me helps their understanding.

5.13 Numerical and visual solutions to 38 × 5

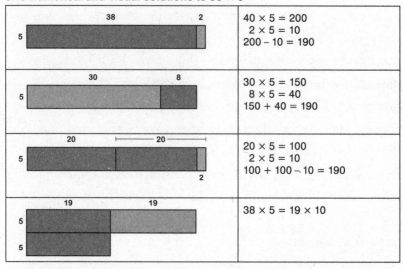

When I showed my Stanford undergraduates the representation of 38 × 5 in figure 5.14, they told me it was the first time they
had understood why you could use this process—of doubling one
number and halving the other—when multiplying numbers.

$$38 \times 5 = 19 \times 10$$

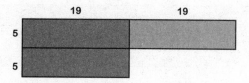

5.14 One solution to 38 x 5

I described the differences between the various ways of calculating 38 × 5 and invited them to consider the different ways to see and think about 38 × 5 as "opening" the question. Some people have been critical of mindset theories, saying that advocates of growth mindset work are putting the responsibility on students to change. I understand that critique, and I believe strongly that it is a teacher's responsibility to open up content so that mindset messages can take hold. If you tell students that they can learn anything but then present narrow content, such as calculating 38 × 5 with one answer and one valued method, students will not see how they can learn and grow. But when the content is opened and students are invited to think and reason, they literally feel their own learning and mindset messages take root and flourish. Although some studies find that mindset messages—given outside of classes, with no change in teaching—have no or little impact,[20] mindset messages given with a change in mathematics approach significantly improve student achievement and beliefs.[21]

In our corresponding visual approach to division, students are invited to build models when they know the total area and one side length.[22] For example, instead of asking students to divide

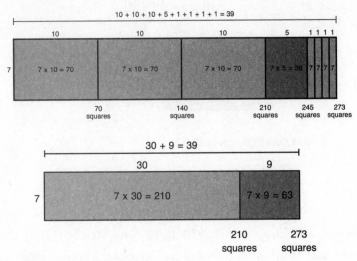

5.15 Different visual solutions to 273 divided by 7

273 by 7, we ask them to construct a rectangle with an area of 273 and one side length of 7, as shown in figure 5.15. Their role is to find the other side's length in different ways.

The visual representations of multiplication and division are important not only because they encourage mental models of the ideas but also because they show the different ways we can think about calculations and the reasons they work. This helps students develop what is known as number sense—the approach to numbers that leads to high achievement.[23] When students have number sense, they can make sense of numbers and use them flexibly in different situations. Number sense does not come from the blind memorization of math facts! Number sense is helped, as I shared in the previous chapter, by asking students to come up with "ish" answers before they calculate, as seen on Mathish.org.

I have been fortunate over the years of my career to work with some incredible elementary teachers who encourage students to develop mental representations by inviting them to think not only visually but also physically. One of those teachers is Jean Maddox, a fifth-grade teacher in the Central Valley of California. Before I met her, she taught multiplication using the district textbook, a page of which is shown in figure 5.16.

This is typical of mathematics books used across the US. Now Jean embraces the notion of mathematical diversity by asking students to think about multiplication—physically, visually, and numerically. Students multiply by building numbers with cubes, by drawing the numbers, and by working numerically. Figure 5.17 shows some examples of different student work.

The students in Jean's class are invited to develop mental models of multiplication that are rich and diverse because they are given opportunities to feel and move physical representations of numbers and to create visuals, words, and number sentences.

Complete to find the product.

1.

			6	4
	X		4	3
+				

2.

			5	7	1
		X		3	8
+					
				.	

Estimate. Then find the product.

3. Estimate:_____   4. Estimate:_____   5. Estimate:_____

       24   37   384
     x 15   x 63   x 45

6. Estimate:_____   7. Estimate:_____   8. Estimate:_____

       28   93   295
     x 22   x 76   x 51

5.16 Textbook approach to multiplication

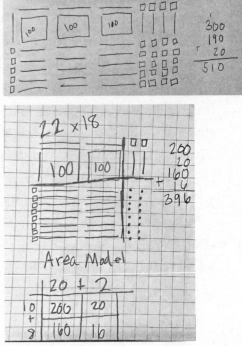

5.17 Examples of student work showing multiplication physically, visually, and numerically

The difference between a narrow task and one that is diverse and creates stimulating classroom environments can be small. For example, instead of asking students to work out the area of a 12 x 2 rectangle, you could ask them how many rectangles they can find with an area of 24 (fig. 5.18). The first question is a calculation; the second explodes into a joyful exploration with visuals, and students are invited to consider the relationship between length and width, giving access to the conceptual understanding of area.

Division of Fractions

One of the most contentious areas of mathematics taught in elementary schools—causing huge difficulties for students, pro-

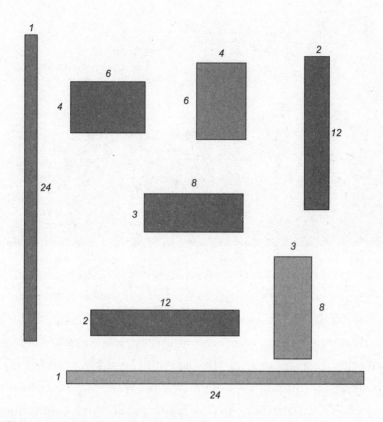

5.18 How many rectangles have an area of 24?

ducing low test scores, and having little practical value—is the division of fractions. When adults are asked to come up with real-life examples of the division of fractions, most cannot think of a single example.[24] Division of fractions, in my view, is a prime candidate for a complete rethink of the way it is taught, including moving it to later years in curriculum standards. Division of fractions is best taught to young children conceptually so that they develop an understanding of the process at work. Instead of being taught with meaning, it is typically taught as a rule: *Divide fractions by flipping and multiplying. In one of the fractions, change the*

nominator into the denominator and then multiply the two fractions. This process is so meaningless for students, it has led to the common refrain: "Ours is not to reason why, just to flip and multiply."

This is unfortunate, as "reasoning why" is the essence of mathematics. Reasoning is at the heart of the subject, and it is the basis for all higher-level mathematical work and mathematical proof. When mathematicians write papers and communicate with each other, they use mathematical reasoning, setting out the logical connections between ideas. When students learn to explain to each other the methods they choose and the ways they use them, they engage in arguably the most important mathematical activity of all—reasoning. I am sure that the phrase "Ours is not to reason why, just to flip and multiply" developed because the operation itself is too difficult to understand or reason about at ten years of age, the age at which the content is taught in the United States.

For the past several years, I have had the pleasure of mentoring and learning from someone I first met as a Stanford undergraduate in my freshman How to Learn Math class. Montse

Cordero came from Costa Rica with a history of very high mathematics achievement in school and in the International Mathematical Olympiad. In Montse I saw someone with genuine mathematical curiosity, whose eyes lit up when we considered why something worked. Since that time Montse has become something of a star among math teachers: Montse appears in my online student

Montse Cordero

course[25] and as a youcubed superhero in our videos for students.[26] But Montse's real stardom comes in their persistence and self-reliance, majoring in math at a school whose department did not ever connect with them as a student or even provide them an adviser, and then pursuing a master's degree in mathematics, fortunately in a different university. Montse is now studying for a doctorate in mathematics.

One of my enduring memories of Montse involves a lesson in which I introduced fraction division to my freshman class and asked them to think about it visually. This was meant to be a brief discussion but ended up taking the whole class period, as the students all realized that they had never understood fraction division; they had only ever "flipped and multiplied." When I showed them a video of Cathy Humphreys teaching seventh-grade students how they might visualize the calculation of 1 divided by 2/3, they were shocked.[27]

In the video you see students make sense of fraction division in three visual ways (fig. 5.19):

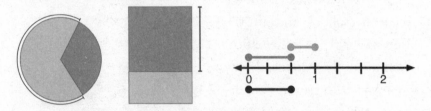

5.19 THREE VISUAL APPROACHES TO 1 DIVIDED BY ⅔

Left, The lighter section represents ⅔ of the circle. Students see that the lighter section fits inside the circle one time plus half of a time.

Middle, The darker section represents ⅔ of the rectangle. Students see that the darker section fits inside the rectangle one time plus half of a time.

Right, The line under the number line represents ⅔. Students see that the lines above the number line show ⅔ plus half of ⅔.

I have a video of me sharing a visual approach to the division of fractions working with a sheep![28] The sheep puppet comes from mathematician Tim Chartier, who likes to help students with mathematics using puppetry.

Years later, in Montse's senior year at Stanford, they decided to study for an undergraduate honor's degree in education—and they chose to focus on the division of fractions. Montse talks about the importance of the moment when they first realized there was a way to understand fraction division in that freshman class and how it had stayed in their mind, and fascinated them, since that moment. Montse's undergraduate thesis starts with this reflection:

Three years ago, I was watching a video of seventh grade students working on a division of fractions problem without using the algorithm they had memorized earlier, and I noticed I didn't truly understand what they were doing.

It was amazing watching 12-year-olds act like tiny mathematicians. They were doing the true mathematical work behind division of fractions, which I had never seen before. I could answer the question using the flip-and-multiply idea I had memorized once upon a time, but I had never even realized that I had no idea why it worked.[29]

Montse took a fascinating approach to their undergraduate honor's study—a computer science perspective—investigating the efficiency and conceptual demands of the flip-and-multiply algorithm and a better algorithm that starts with finding common denominators.[30] One of Montse's interesting findings is that to understand the flip-and-multiply algorithm, students must be given content that they have not yet learned in the grade level in which division of fractions appears![31] This is important as it suggests that *we are expecting students* to learn and use the algorithm, to just accept it, without understanding it, without "reasoning why."

There is a problem with introducing algorithms that students do not understand. Elementary teachers do important work helping students to develop number sense and, more generally, to make sense of the mathematical ideas they meet, using important visual and physical representations, but when students are taught algorithms, it seems that their sense-making stops. Pesek and Kirshner studied this phenomenon of learning rules followed by an inability to think in different ways, calling it "cognitive interference."[32] When students first learned algorithms and then learned conceptually, engaging in mathematics in diverse ways, they did not do as well as the students who only learned conceptually. The learning of methods and rules seemed to *interfere* with the students' development of helpful mental models. I will return to this study, and share more details, in chapter 6.

Students have problems with not only the division of fractions but also the addition of fractions, as shown by a National Assessment of Education Progress (NAEP) question given to thirteen-year-olds across the US. When these students were asked to estimate the answer to $^{12}/_{13} + ^{7}/_8$ with the choice of answers of 1, 2, 19, or 21, the most common answer was 19, followed by 21. Their answers came from adding the numerators (19) or adding the denominators (21). Only 24 percent of students chose the correct answer of 2.[33] If the students had received opportunities to take an ish approach to number questions, they might have seen that the estimated sum of the two fractions is 2. As this example illustrates, taking an ish approach would help students in many standardized test questions.

Why am I not surprised that students give nonsensical answers? When students learn fractions, they do not think conceptually or in an ish way about them; they more typically learn a set of rules:

- To add fractions, make common denominators and add the numerators.
- To subtract fractions, make common denominators, then subtract the numerators.
- To multiply fractions, multiply the numerators and the denominators.
- To divide fractions, change the division sign to multiplication and change the numerator of one of the fractions to be the denominator and the denominator to be the numerator (flip and multiply).

All of these rules trample over the most important part of fraction understanding: the fact that the two numbers—the numerator and denominator—are in a relationship. In thinking about fractions, it does not matter what size the numerator or denominator is; it matters how big they are in relation to each other. When we ask students to act on the numerator or on the denominator, they forget what the fraction means. When students add $12/13 + 7/8$ and answer 19, they are not thinking about the relationship between the numbers 12 and 13 or 7 and 8. This is because they have not spent time considering fractions as a whole, making sense of the numerator and denominator *in relation* to each other; they have not developed mental models of the fractions, and they have not given ish answers.

I still remember the evening I spent with my eldest daughter who was in fifth grade at the time. She had been struggling with her school's approach to fraction rules, and her teacher told me she just "could not" understand fractions. This may have been, in part, due to her learning differences, which some teachers in her life have decided are deficits rather than differences. (Fortunately, she also had some wonderful teachers who knew thinking differently did not mean thinking inadequately.) She and I decided to look at fractions together that night, as she was taking a test the next day. I did not have much time with her compared to the many hours she had spent studying fractions in the classroom, but in one hour I did this: I taught her that she should always look at the relationship. Together we visualized different fractions and considered their meaning, the relationship between the numerator and the denominator. Before working on any operations, we looked at the overall value of the fraction. Then we tried adding, subtracting, multiplying, and dividing, taking an ish approach to the fractions, thinking about what the approximate answer would be, before working out the exact answer. I encouraged her to consider the big picture and the focused visual models that are so important to learning (which I discussed in the previous chapter). She came home from school the next day beaming from ear to ear, telling me she got the highest score in her class on her fraction test. Part of my free online student course, taken by over one million people, shares this relational, ish approach to fractions, encouraging participants to build mental models.[34]

I am making a case for students taking a different approach to fractions at home and in school, one that is based on sense-making, number relationships, and visual and physical thinking that leads to mental representations. I am not arguing that algorithms are not useful; I am arguing that they should not be taught to students until they understand fractions conceptually. Students need to understand what a fraction is and be given a lot of experience thinking about the value of fractions as a conceptual whole.

Fraction division could be the most hated part of the math curriculum, the area most likely to lead adults at parties I've attended to groan with pain—although algebra is another contender for the title of Most Hated. But this dislike can be transformed when we give students opportunities to develop visual and physical models that ground their understanding. For example, if you were asked to calculate 1 divided by $3/4$, you could flip and multiply, arriving at the correct answer, $1/1 \times 4/3 = 4/3$, but you would have little or no idea what was happening.

Or you could sketch the fraction, asking a different question: How many times does $3/4$ fit inside 1?

Note in figure 5.20 that ¾ goes into one whole, one time with some left over. If you imagine the ¾ fitting into the empty space, you will see that only ⅓ of that piece fits. This gives us the

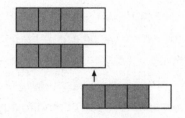

5.20 Visual approach to 1 divided by ¾

answer of 1 and ⅓. Three-quarters fits inside one whole, one and one-third times. Thinking about the fraction-division process visually, and with an ish lens, may take a little longer than flipping and multiplying, but it provides a mental representation that can anchor any future work with calculations and algorithms.

For division of a fraction by a fraction (something most people have never used in their lives after school), I prefer to make both fractions have the same denominator first. Changing fractions with different denominators to have the same denominator is work that focuses on the relationship, which is valuable for students. Let's consider ¾ divided by ⅔. We can give both fractions a denominator of 12:

¾ × 2 = ⁶/₈ ¾ × 3 = ⁹/₁₂	⅔ × 2 = ⁴/₆ ⅔ × 3 = ⁶/₉ ⅔ × 4 = ⁸/₁₂

Multiply each fraction (which keeps the same fraction) until they both have the same denominator, in this case 12.

We have now changed the question from ¾ divided by ⅔ to ⁹/₁₂ divided by ⁸/₁₂. Thinking about the first and second questions with an ish approach, you will probably find it difficult to estimate ¾ divided by ⅔ but can pretty easily figure out how many times ⁸/₁₂ goes into ⁹/₁₂: our ish answer should be a little over 1.

Teaching this to students, I would start by drawing ⁹/₁₂ divided by ⁸/₁₂.

Figure 5.21 shows that 8/12 goes into the space 1 and 1/8 times.

This visual approach to mathematics can extend to any grade, any level, any mathematics, and any task.

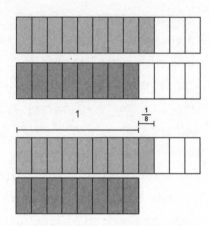

5.21 Visual approach to 9/12 divided by 8/12

Algebra

It was my and my team's first time running a summer camp for students since learning the evidence about brain plasticity and mindset. The previous five years had produced extensive data showing that there is no such thing as a "math brain," that all of our brains are growing, strengthening, and connecting all the time.[35] Research had shown how important it is to believe in your own potential, to have a "growth mindset."[36] We gathered eighty-two students from a local school district who had not had good math experiences. They were mixed in their prior achievement, and they were culturally and racially diverse. Some of the students scored zero on the test given to them in their school district before they came to us; others scored at very high levels; and still others were at levels in between. The students came from two different grade levels; that summer they were preparing to enter seventh or eighth grade.[37] We decided that the most useful content to teach them would be algebra. In teaching this critical subject, we knew it would be important for students to understand the concept of variability and the meaning of a variable. I chose one of my favorite problems, which starts by asking

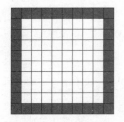

5.22 How many squares are on the border of a 10 x 10 square?

how many squares are tiled on the outside edge of a 10 x 10 square. I showed them a visual like the one in figure 5.22, just for a few seconds so they would not count but would group the numbers.

Students shared many different answers, including 36, 37, 38, 40, and 44, which led to a spirited conversation. I told them I loved that there were different answers, as that meant it was an interesting problem and we would have lots to think and talk about. After chatting for a while, the class decided together that the total number of squares on the outside edge of the square was 36. As each student shared their way of seeing 36, I drew it on the dry-erase board, so we had an array of visuals, as shown in figure 5.23.

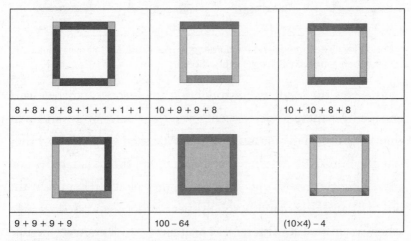

8 + 8 + 8 + 8 + 1 + 1 + 1 + 1	10 + 9 + 9 + 8	10 + 10 + 8 + 8
9 + 9 + 9 + 9	100 − 64	(10×4) − 4

5.23 How many squares are on the border of a 10 x 10 square? Shown visually and with numbers.

We continued working on different-size squares over a couple of lessons, sharing with the students that we could use a variable to represent the pattern growth, so our thinking about the total numbers could apply to any size of square. All the different ways students saw the tiling pattern can be expressed as equivalent algebraic expressions, as shown in figure 5.24:

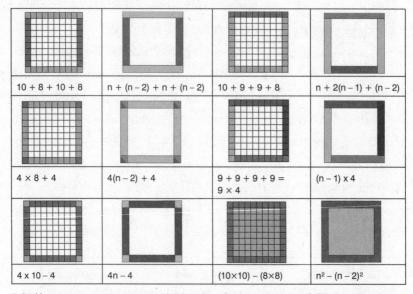

$10 + 8 + 10 + 8$	$n + (n-2) + n + (n-2)$	$10 + 9 + 9 + 8$	$n + 2(n-1) + (n-2)$
$4 \times 8 + 4$	$4(n-2) + 4$	$9 + 9 + 9 + 9 = 9 \times 4$	$(n-1) \times 4$
$4 \times 10 - 4$	$4n - 4$	$(10 \times 10) - (8 \times 8)$	$n^2 - (n-2)^2$

5.24 How many squares are on the border of a 10 x 10 square? Shown visually, with numbers, and with algebraic expressions.

This was the students' introduction to algebra, and they were engaged by the activity—seeing what variables were and why they were useful in communication. We asked students how they saw the border of the shape, and I recorded their visual ideas on the whiteboard, with the students' names next to them. On the board it said, "Josh's method," "Elisa's method," and so on. Later, when we moved to different-size squares, students were invited to choose a method—such as Elisa's method—and investigate it. As students worked, they discussed their different ways of seeing and thinking, collaborating with each other, and building on each other's ideas. Importantly, the students' conceptions of algebraic generalization were visual—and beautiful. At the end we had six different expressions of the same mathematical function (4n − 4), all of them equivalent and all of them illuminated by the visual representation.

When students use algebra to describe patterns, they are using variables as a language to describe the world—which is an important and helpful use of variables that have more meaning than endless examples of solving for x.[38] They are also developing mental models of algebra and of generalization, as they can see the pattern growth. Students need this important introduction to algebra, and we share it through four free weeks of lessons on youcubed.[39]

One of the most beautiful examples of visual representations of algebra comes from a woman I had the pleasure of teaching in Stanford's teacher education program. In my class preparing people to become mathematics teachers, when I share the idea that mathematics can be taught as a diverse subject, with visuals and other mental representations, sometimes a spark ignites in those who hear it, and they extend these ideas to many areas of their lives. This was the case with Diarra Buosso, who grew up in Senegal and now uses linear and quadratic equations to create beautiful clothing that is sold in leading stores worldwide (Nordstrom, Shopbop, Stitch Fix, and many others).[40] She accomplishes this impressive feat while teaching high school mathematics in California.

Diarra has been interviewed by many journalists; her story has been shared in *Vogue* magazine, on CNN, and in other news outlets.[41] In different interviews Diarra shares that she comes from a long line of artisans and craftspeople in Senegal, but she loved math and traveled to the US for college before becoming a Wall Street analyst. During those days, Diarra shares, she felt deeply unfulfilled because she loved both art and math and could not choose between them. She did not want to be calculating all day, without any artistic expression, nor did she want to work in an artistic field, without any mathematics. Her desire to do something different from her banking work led Diarra to my teacher education class at Stanford.

During our year together, Diarra learned about visual, creative, diverse versions of mathematics, and she was enchanted. One day after class she pulled me aside and shared that she was interested in using mathematical formulas to design clothing. She said she expected me to dismiss the idea as being "too nerdy" for anyone to be interested. Of course, I did not dismiss the idea; I told Diarra it sounded amazing. Diarra went on to create beautiful clothing designed from linear and quadratic functions; examples are given in figure 5.25.

Diarra says that the year in our class helped her understand she could pursue art and math together, which unlocked her creativity and led to deep personal fulfillment.[42] She is an amazing teacher who excites her high school students, encouraging them to turn algebraic expressions into art (fig. 5.26). To connect with her students, she asks them how they spend most of their time; her students report spending an average of twenty-six hours a week on social media, so she meets them there. After getting approval from her principal, she opened an Instagram account and uses Instagram Stories to share polls and questions with students. Ninety-two percent of them report that this is helpful for their learning. They appreciate being able to vote on answers and enjoy getting shout-outs for their work. According to Diarra, "Gamifying math through a platform kids already associate with play and fun made the subject more accessible."

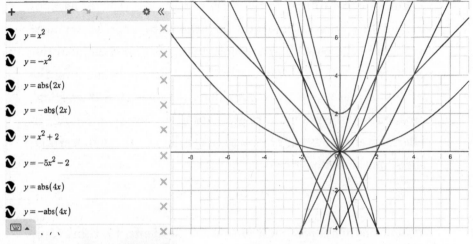

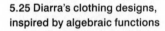

5.25 Diarra's clothing designs, inspired by algebraic functions

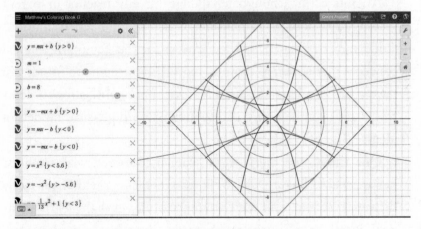

5.26 Example of student work using Desmos graphing software

Diarra came to realize that her students didn't hate math, as they first told her; they just disliked the ways it had been taught to them in the past. Diarra is now an activist for the global celebration of African culture and is passionate about merging the worlds of math and art to provide students with a diverse approach to mathematics. I am proud to have played a small role in Diarra's journey, as her creativity was unlocked through learning about mathematical diversity.[43]

When mathematics questions invite students to think only numerically, important opportunities are missed—for brain connections, for increased access to understanding, for deeper engagement and motivation, and for development of critical mental models. I hope to show in this chapter that it is not difficult to make a question visual; it can often be as simple as asking people, *How do you see it?* When students are asked this question and share their ideas, they feel ownership of the mathematics as they develop important mental models.

An interesting thought challenge for any teacher or parent reading this book is to consider which mental representations students are being invited to develop for the key mathematical concepts that make up their domain. If none come to mind, you have a fun challenge ahead: creating opportunities for students to visualize mathematics and to interact with physical models of ideas.

One traditional high school teacher I worked with changed her approach. She used to walk into lessons, provide a perfectly delivered lecture, and have students work through nearly identical questions. She told me that this all changed one day when she began by simply turning to her class and asking them what they thought about their math questions. She was amazed that her students happily shared different ideas, and they engaged together in meaningful conversations about the questions, considering how the different methods worked and drawing visuals on the front board. After this she never went back to her perfectly delivered lectures.

Likewise, in my own teaching, I honor students' thinking, and their desire to be engaged as social creatures, by asking them how they see and understand ideas. I know that my greatest resource in the classroom is student thinking.

My favorite lesson starter is to share a visual representation of mathematics and ask students, *What do you notice? What do you wonder?* These inviting words, first suggested by Annie Fetter, can be used when we share dots in dot card number talks; when we share data talks, inviting students to make sense of real-world data and encouraging their development of data literacy; and when we share geometric drawings.[44] Examples of all three are shown in figure 5.27. (In chapter 6, I will share what happened when I was invited to a First Nations school on

5.27 Start the day with a dot talk, a data talk, or a shape talk

a reserve in Canada, whose students and teachers are part of the Okanagan Nation, and we explored a visual of a dream catcher together. The student discussions, which led to visual representations of algebraic functions, were incredible.)

The great benefit of starting classes, or family conversations, with a visual invitation is that students are encouraged to share their developing ideas, in which there are no right or wrong answers and the classroom or family space is open to diverse perspectives. Instead of the beginning of class being a dreaded review of homework, teachers can start with fun and interactive questions, and students can share and build on ideas, opening their minds—and building confidence—for the mathematical adventures that follow!

6

THE BEAUTY OF MATHEMATICAL CONCEPTS AND CONNECTIONS

Mathematics is a conceptual subject. Many people believe it is a list of rules and methods, but it is actually a small set of important concepts that students—and all people—can meet, understand, and love. Two researchers at the University of Warwick in England, Eddie Gray and David Tall, showed this powerfully in their 1994 study of children's approaches to arithmetic and numbers.[1] I utilized their study for many years, but recently I was able to replicate it with young learners in San Jose, California. The reason I find their work so important is that they uncovered the difference in the behavior of high- and low-achieving students. The study we conducted in the US, some twenty-nine years later, found the same result as well as a new insight.

Gray and Tall gave students between the ages of seven and thirteen a series of arithmetic problems, such as 6 + 19, and recorded students' strategies for solving. The researchers had previously asked teachers to nominate both low- and high-achieving students. They found something fascinating—the high achievers used number sense to answer the questions. Number sense is considered to be in place when people can approach numbers flexibly: for example, breaking them apart in different ways, seeing them visually, and using different strategies to act on them. In the study, when tackling 6 + 19, a student with number sense added 5 and 20 instead. The low-achieving students, by contrast, did not take this flexible approach to numbers. Instead, they used counting methods, laboriously adding the single digits. For example, when they were given 19 − 16 they "counted back," starting at 19 and counting down 16 numbers, which is really difficult. The students who used number sense did something much easier, subtracting 10 from 10 and 6 from 9.

The researchers drew an important conclusion from their study: low-achieving students are often low achieving because they are not approaching numbers conceptually and flexibly but as methods and rules; they have not learned to engage with numbers flexibly. But when districts and schools realize that students are low achieving, they often pull them out of class and give them worksheets full of "drill and practice," cementing their approach to arithmetic as a set of rules. This is the opposite of what most people need.

NUMBER SENSE IS THE KEY

In figure 6.1, Gray and Tall highlight the important difference between regarding mathematics as methods or rules and regarding mathematics as concepts or ideas when students are learning numbers and arithmetic.

Gray and Tall share that when students learn to count, they are learning a method, but that should lead into an understanding of the concept of a number. When they learn the method of

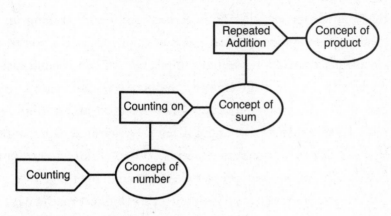

6.1 Mathematics is a conceptual subject
Gray and Tall

counting on, it should lead into a concept of a sum, and when they learn the method of addition, it should lead into the concept of a product. Learning concepts involves thinking deeply, considering, for example, *What is a number? How can it be broken apart in different ways to make other numbers? How can it be represented visually? Where do we see numbers in the world?* Some students never learn to think conceptually because the teaching they experience is all about rules and methods.

If you learned math (or anything) as a set of rules, it is not too late to take a conceptual approach now, which can give you a whole different insight into the important ideas that make up our world. I have met many adults who were not taught to approach numbers flexibly and wished to start doing so.

One of the many problems with a rule-based approach is related to an interesting brain process called compression. When we learn new knowledge, it takes up a large space in our brains—actual physical space, as the brain works out how it fits with other knowledge we have. When young children learn addition for the first time, it takes up a large space in their brains. Over the years, that knowledge of addition becomes compressed, taking up a smaller and smaller physical space in their brains. As adults, if we are asked to add 3 + 4, for example, we can quickly and easily retrieve that knowledge from its compressed small space in our brains. This compression makes space in our brains for more and more learning. But Gray and Tall, in their seminal paper, argue that we can only compress concepts. When children learn only rules and methods, compression is not even happening.[2]

William Thurston, who won one of the highest mathematical honors, the Fields Medal, for his work, wrote about compression in this way:

> Mathematics is amazingly compressible: you may struggle a long time, step by step, to work through the same process or idea from several approaches. But once you really understand it and have the mental perspective to see it as a whole, there is often a tremendous mental compression. You can file it away, recall it quickly and completely when you need it, and use it as just one step in some other mental process. The insight that goes with this compression is one of the real joys of mathematics.[3]

Thurston describes compression as the reason for his joy in learning mathematics, and yet, as Gray and Tall point out, students learning mathematics as facts and rules never get to experience this important brain process. The fact that they never learn mathematics conceptually may be the reason they rarely describe mathematics as "a real joy."

When, in 2023, my team at Stanford repeated the Gray and Tall study to see whether their results still held (twenty-nine years later and in a different country), we asked teachers of grades one to five to nominate their three highest- and three lowest-achieving students in each class. We were particularly interested to find out if the two groups differed in their approach to mathematical questions, as Gray and Tall discovered. In our study, we collaborated with neuroscientist Bruce McCandliss and his team to add questions on groupitizing (explained in chapter 5), a concept that was not known at the time of Gray and Tall's study. We gave the thirty students a groupitizing assessment, and then we gave them six arithmetic questions. The students sat with an interviewer, one on one, in a quiet space with a small table and chairs. A collection of counting chips sat on the table, and we told the students they should feel free to use them at any time. The students' responses to the questions were then double-coded by a team of researchers.

Our new study had two significant findings, one of which replicated Gray and Tall's: the students who were high achieving approached the arithmetic problems with number sense; the students who were low achieving used less effective counting methods. Our study also found that the students who were

groupitizing were more likely to use number sense and to be high achieving. This was the first link between groupitizing and number sense ever found.[4]

Our study showed, like Gray and Tall's, that students were high achieving not because they knew more but because they approached numbers differently. Their approach differed from the low-achieving students' in that they approached numbers conceptually and flexibly—breaking numbers apart and making different numbers to solve problems. As groupitizing is about breaking numbers up in different ways, as chapter 5 explains, it is not surprising that students who learned to groupitize also learned to develop number flexibility.

The conceptual approach to numbers that the high achievers used, and that we all can use, often involves stepping back from detailed methods, such as adding, and focusing on a bigger idea or concept. One way we can help all students develop a conceptual and flexible approach to numbers is to ask them to consider numbers conceptually and take an "ish" approach to their work.

THE PROBLEM WITH STANDARDS

We were in the middle of the global COVID-19 pandemic when I received a call from an education leader working for the State of California. She asked if I would help teachers during the lockdown by highlighting the more important mathematical content students should learn. The State Board of Education had recognized that the pandemic was making teaching an even more difficult job than usual, and teachers simply could not get through all the individual, isolated standards they were meant to be teaching. I believe that, no matter the circumstances, it is *always* impossible for teachers to teach the vast amount of content they are expected to cover at anything but a surface level, and I have never met a mathematics teacher who does not think there is too much to teach in their grade level or course.

I was already one of the writers for the new state framework when I received the call, and I was still doing my regular work as a professor and the codirector of our youcubed center, so I myself was under a lot of time pressure. Despite this, I agreed to help, as it seemed like an ideal opportunity to help California teachers move to a connected and conceptual approach.

In the past, organizations have tried to help teachers approach the excessive amount of content set out in standards by organizing them by importance. I did not take this approach. Instead, I worked with my colleague Cathy Williams, cofounder of youcubed, to reorganize the mathematics standards into a set of connected, coherent, big ideas.[5] Figure 6.2 shows examples from kindergarten and grade 6.

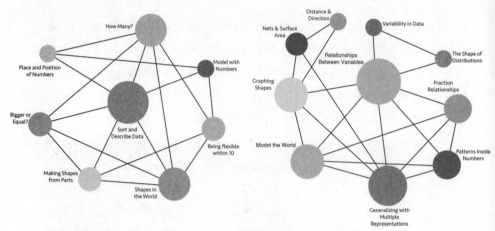

6.2 Mathematics as connected, big ideas in kindergarten and grade 6

Mathematics committees across the state reviewed our reorganization of content into big ideas, and it was unanimously approved by the California State Board of Education in May 2021. This approach of teaching mathematics through big ideas was also a central part of the new California Mathematics Framework, which was unanimously approved in July 2023.

When teachers focus on bigger ideas rather than small methods, students can access conceptual understanding, and the smaller methods can be learned inside the big ideas. The National Research Council is an important body within the National Academy of Sciences that issues advice to inform policy within and outside education. When a group of scientists was charged with communicating research on learning in a book for educators, they gave the following advice:

> Superficial coverage of all topics in a subject area must be replaced with in-depth coverage of fewer topics that allows key concepts in the discipline to be understood. The goal of coverage need not be abandoned entirely, of

course. But there must be a sufficient number of cases of in-depth study to allow students to grasp the defining concepts in specific domains within a discipline.[6]

The National Academy of Sciences' call to replace superficial coverage with a small number of cases of in-depth study was part of our motivation for turning all the disconnected mathematics standards into a set of big ideas. The beauty of teaching and learning a smaller set of connected, bigger ideas is that teachers and parents have more time to go into depth on each idea, and they allow students to think conceptually. When students can dive deeply into the mathematical concepts, they learn the same mathematics, but instead of learning disconnected methods piece by piece, they learn a set of connected ideas and methods through rich tasks. This approach was set out both in the 2023 California Mathematics Framework[7] and in our K–8 books.[8]

If students learn, deeply, the eight or so big ideas at their grade level, those ideas will serve as a foundation for everything else they learn. When teachers focus on the big ideas using rich, deep

tasks through which students explore the mathematical concepts deeply, the smaller methods are often learned inside them, but now with meaning.

Neuroscientist Jeff Hawkins's proposal about how the brain holds knowledge and information gels with the foregoing educational advice. He argues that the brain arranges all knowledge using reference frames—these are located in our brains, as big cities might be located on a map.[9] As mathematicians work, we need reference frames to know where we are and what we need to do to move from one area of knowledge to another. Often these reference frames have visuals that we can see or imagine in our minds. Hawkins gives the example of DNA. If you have studied genetics, when someone talks about DNA, your brain is likely to pull up a visual of a double helix that can unzip. You have never seen a DNA molecule, but your brain has made a reference frame for this area of knowledge, which helps with organization.

If people are learning effectively, they are making reference frames. They are not just storing facts, one on top of another; they are relating their knowledge to bigger, conceptual reference frames. Hopefully, these reference frames have mental representations to ground them. Our recommendation is to help students learn a set of big ideas deeply and well, so that these big ideas serve as reference frames that can organize all the students' knowledge.

When teachers ask me to help them teach well, even with long lists of content to teach, mandated tests, and disconnected textbook questions, I give this advice: make sure each of the eight or so big ideas in your grade level are learned through rich, deep activities, because if these foundational ideas are learned deeply and conceptually, they will provide a structure for all the rest of students' learning. I would give this same advice to anyone learning any new content. Instead of focusing on small details, consider the bigger picture: What do the details really mean? And how are they connected to each other? Sometimes the important concepts are hidden behind the details that are set out.

Students who are asked to learn and remember terms in any subject would do well to take this conceptual approach to their content. Cathy Williams is the cofounder and codirector of youcubed, but prior to this and her work as a district and county leader, she was a high school teacher who taught both mathematics and AVID—a program that prepares students, particularly the most vulnerable, for a college future. She recalls that students would come to her confused and overwhelmed by the different terms they were given to learn. When they told her they had to remember *mitochondria, nucleus, ribosome,* and *cytoplasm,* she invited them to write and visualize a story that was meaningful to them that connected the terms in some way. This helped the students engage with the ideas and move from an approach of memorization and detailed focus to conceptualization and big-picture thinking.[10]

MATHEMATICAL CONNECTIONS

Another goal in our work for the state was to share mathematics as a set of connected ideas. If you ask any thoughtful mathematician to describe their subject, they will describe a small set of important ideas, with many rich connections, similar to the network maps in figure 6.2 (page 182), and similar to the reference frames described by neuroscientist Jeff Hawkins. But when writers of standards set out content, they take the interconnected map of ideas and chop it into tiny pieces, which are then shared with teachers as hundreds of curriculum standards. It is no wonder students regard mathematics as a disconnected subject; that is the way mathematics is presented to teachers. The same disconnected, small standards are used by textbook authors to make questions. When we set out the big ideas in mathematics for the state, we shared the connections between the ideas—bringing back the relationships between different concepts, which show the ways ideas can build from and with each other.

Jim Stigler, a psychologist at UCLA, studies the ways experts hold knowledge. He and his team highlight the importance of conceptual understanding, mental representations, and the connections between mental representations for the development of expertise.[11] They point out that when people develop connected understanding, they are more effective at solving problems.

When people recall and attempt to implement knowledge that is not connected to other concepts, such as the relationship between a part and a whole, the problem-solving stops and they give up. By contrast, someone with connected knowledge

who is asked a question about ratio and remembers what they learned about operations will be able to connect their knowledge and follow a path to solve the ratio question.

Sarah Nolan is a fourth- and fifth-grade teacher in California who was part of a group of teachers working with me and my team at

Stanford. Sarah was inspired by the idea of teaching mathematics as a conceptual and connected subject. She started teaching from our big idea books and explained to her class that they would look for the connections between ideas together.[12] She told them that math is a conceptual web, and together they would see how the threads between concepts helped them to understand other concepts. Sarah set out the mathematical topics as big ideas for the year on her classroom wall and modeled a way of pointing out connections between them. The students then took over the process, adding their own connections whenever they saw them. This happened organically—when students saw connections, they would write them on a piece of paper, which the class pinned to a rope that connected ideas on the wall. A photograph of the connections wall, taken midway through the year, is shown in figure 6.3, and a video of Sarah's classroom including the connection maps can be found on youcubed.org, called "moving from maths anxiety." By the end of the year, the wall was like a spider's web. When Sarah took the students' connections down, those who had contributed them asked if they could take them home—they were proud of their connection-finding!

Another of my favorite methods for helping people develop conceptual and connected knowledge of ideas is making sketchnotes.[13] A team of researchers from Spain and Australia describe sketchnotes as "a visual thinking form that integrates notes and sketches to explain scientific topics."[14] Two teachers have been kind enough to share sketchnotes they have made showing some of my teaching ideas (fig. 6.4).

People are increasingly using sketchnotes to represent and remember ideas shared in conferences, workshops, and business

6.3 Mathematical connections in Sarah Nolan's classroom

meetings, but teachers can also show students how to use them to highlight big ideas and their connections, a focus that all learners should develop, whatever they are learning.[15] I particularly love sketchnotes because their form, much like concept maps, can communicate ideas in ways that traditional methods

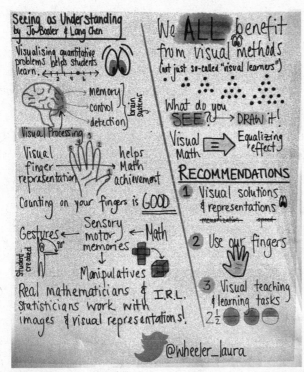

6.4 Two sketchnotes sharing my teaching ideas
Laura Wheeler and Impact Wales

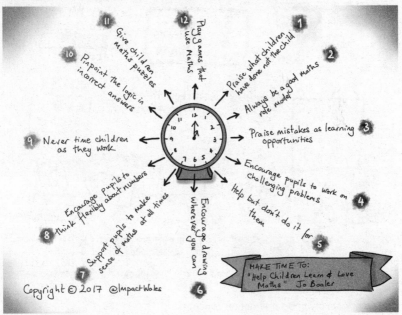

did not or could not. The previous chapter shared research on the value of seeing mathematics visually, and sketchnotes are a great way for students to make their own visuals.

Janet Nuzzie, a leader in mathematics education, works to help students learn mathematics with depth, understanding, and mindset ideas. She and I sketched some of the most important concepts in teaching fractions, which are shown in figure 6.5.

Research has shown that when students take notes with laptops, they retain less than when they sketch or write ideas by hand.[16] Traditional note-taking—writing ideas down verbatim—involves shallow processing, whereas creating sketchnotes requires students to process the information,

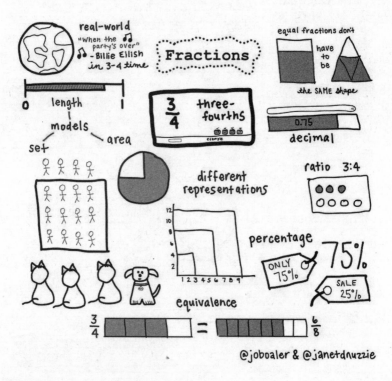

6.5 Sketchnote of important fraction concepts

consider the bigger picture, visualize it, and reframe it—all valuable learning acts. Research has even found that students making sketchnotes of ideas increased their achievement on mathematics word problem–solving and increased their engagement and motivation—this was particularly impactful for students with learning differences.[17] Heidee Vincent, a university mathematics professor, uses sketchnotes to help her students understand mathematics; her sketchnote showing important ideas in data literacy, statistics, and data science is shown in figure 6.6.[18]

I encourage you to try making a sketchnote of ideas you are working with in your life. If you feel hesitant because you believe that you cannot draw, consider using icons that are provided for free on the internet.[19] Remind yourself (and your students) that the main goal of a sketchnote is not to produce a beautiful piece

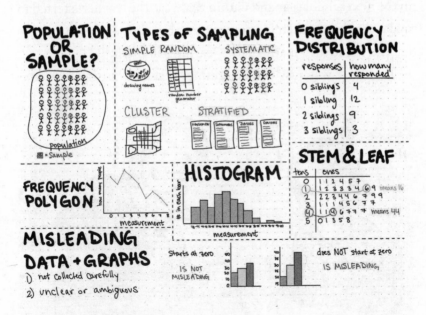

6.6 Heidee Vincent's sketchnote

of art but to represent ideas, give the ideas personal meaning, and think conceptually, making connections between ideas.

TEACH CONCEPTS AND CONNECTIONS

I remember the day when Shelah Feldstein, the math director from the county office of Tulare, California, found her way to my office at Stanford.

She asked me if I would help the county leaders bring the approach we were recommending—of students working on big ideas and connections—to fifth-grade teachers in the region. Together, we formed a plan for the teachers to take my online math education course and then meet in groups to plan classroom changes based on the new ideas they were learning.[20] We started in the next school year and within that year the teachers started to make changes in their classrooms. As they did so, they started to see their students' mathematics learning flourish.[21]

That year was just the beginning; teachers in the area con-

tinued to engage in professional learning, while county leaders spread the new ideas to other teachers. Teachers across grade levels began to learn this different approach to teaching.

A few years after that initial meeting in my office, I sat down with some of the fourth-grade teachers, Annie Braun, Jeremy Kemp, and Stephanie Gomes. They were excited to discuss the ways they had changed as teachers—encouraging students to think in new ways and to celebrate mistakes—but there was one aspect of their change that interested me in particular.

Jeremy described an "eye-opening aha." He had engaged his students in a rich, deep, and extended fraction task but was getting nervous about the other content in the curriculum he had to teach—decimals, geometry, measurement, and other topics. His nervousness quickly dissipated when he led students on to decimals:

> The kids had spent so much time on fractions, learning the ins and outs, they could take their knowledge from that concept and apply it to other aspects of math. When we switched to decimals, I was blown away with their ease in the concepts—they were able to take what they had learned in fractions and without even practicing it they could apply it to decimals. It was amazing.

Annie and Stephanie agreed, adding that it was rare for students to see mathematical connections, which they all thought was "really cool and amazing." At the end of the first year of our work with the leaders and teachers in the Central Valley, the students who had been on their learning journey achieved at significantly higher levels in state mathematics tests—especially girls, language learners (sometimes referred to as English language learners [ELL] or English as a second language [ESL] students), and students from lower-income homes—all of whom typically underachieve in mathematics.[22] After the county office ran its well-thought-out professional development with the teachers, the students in Tulare County started to experience mathematics as a conceptual, connected, and diverse subject.

SUCCESS IS LINKED TO CONCEPTUAL TEACHING

Not all students experience such meaningful mathematics instruction; I have witnessed many times the problems students run into when they have not been taught to approach mathematics as a conceptual and connected subject. On one of those occasions, I was watching a great teacher—Cathy Humphreys— teach a group of middle school students in California, all of whom were designated as language learners.[23] Cathy had been brought in to teach a unit on fractions; she had been told by the regular teacher that the students were having difficulties learning the subject. Cathy planned lessons that invited students to learn fractions visually and conceptually. During one of the lessons, Cathy showed them the visual in figure 6.7 and asked the class a question that she thought would be a simple one to answer: "What fraction of this shape is shaded?"

Cathy quickly noticed that few students were raising their hands, and all of them seemed unsure, so she invited

6.7 One quarter

them to discuss the problem in groups. This is when I heard one of the most fascinating and enlightening mathematical discussions I have ever listened to. Here is part of it:

HUGO: I think it is one two'th—there are two shapes, and one is shaded.

LUCAS: I think it is one third.

SOFIA: I think it is one whole.

HUGO: No, it is half—one out of two pieces.

PABLO: I think it is one fourth.

LUCAS: Why?

PABLO: Well, if you were to imagine lines right there and right there *(he divides the shape, showing with a pencil in the air lines dividing it horizontally and vertically)* . . .

LUCAS: But there's no lines right there. Plus, she didn't say, "Can you prove," she just said, "What fraction is right there?"

I was not surprised to hear students say the answer was one half or one third, as these are common misconceptions. What was so interesting to me was that when Pablo gave a clear and correct explanation for why the shape was one fourth, showing the lines that could be drawn to divide the shape into four equal pieces, Lucas objected, noting the absence of visible lines.

What we witnessed in this interaction was a student following what he thought were the rules of math class—you answer the question asked; you do not change the question, by adding lines, for example. But what Pablo was doing is one of the most important mathematical acts a student can learn—it is called shape flexibility. Just as number flexibility is important, allowing us to change numbers into friendlier numbers when calculating, shape flexibility allows us to move parts of a shape, or add lines, to increase understanding. Michael Battista refers to this as spatial structuring,[24] and it is intricately tied to number sense.[25] Lucas demonstrated his close attention to the teacher's language, arguing that perhaps they could have taken Pablo's flexible approach if they had been asked to *prove* their answer rather than just state the fraction. Both of these responses indicated to me that Lucas had learned to follow rules, and his learning of rules was interfering with his ability to consider Pablo's accurate mathematical reasoning.

The lesson got even more interesting when Cathy asked the class to share their discussions. Jesús walked to the board and gave a beautifully clear explanation, showing the shape divided into four parts and saying,

> I think it is one fourth because if you divide right here and right here *(he divides the square on the board into four)*, you get four squares, and they all have the same area, and if you shade one in, there would be one fourth.

Cathy asked the class if Jesús convinced them, saying that if he did not, they should ask Jesús a question that would convince them; she invited the class to "make him accountable for his proof." This is when Jorge asked a stunning question of Jesús. He asked Jesús: "What are the rules?"

The class waited in suspense. Jesús, at the board, shrugged and looked unsure. Jorge continued, asking, "How do you know you are right, if you don't know the rules?"

These moments are fascinating to me, as they illustrate, through a class discussion, what researchers have termed "cognitive interference."[26] Jorge and Lucas, and probably many oth-

ers in the class, had been taught fraction rules. These rules had taken such a firm place in their minds that they were blocking their ability to think conceptually and to consider the mathematical reasoning that is key to their understanding. It was as if they did not believe that mathematical reasoning and sense-making were legitimate acts; what they thought they needed to do was follow rules.

A similar process happens when young students are taught algorithms in their early learning of arithmetic and fractions. I have seen teachers do important work developing number sense in their students, but when they teach the students algorithms, it is as though all the students' sense-making disappears, and they start to blindly follow algorithms. (I discussed this in chapter 5 in relation to fractions.) I am not opposed to the teaching of algorithms, but I do believe that students are often introduced to them too early, before they can reasonably understand them. This pushes students into rule-following mode and seems to point them away from the conceptual thinking that is so important.

Dolores Pesek and David Kirshner, researchers in mathematics education, became interested in the concept of cognitive interference after observing multiple cases of students being hindered in their mathematical thinking by rules they had been taught.[27] One of the cases involved the learning of algebra,[28] a second involved the learning of decimals,[29] and a third involved the learning of fractions.[30] These different cases prompted Pesek and Kirshner to carefully investigate the idea of cognitive interference, which they define as a problem that occurs when "previous understandings in a domain are so powerful as to spontaneously obtrude into subsequent learning."[31]

To investigate this phenomenon, which they speculated could be the root of many students' mathematical problems, they designed a controlled study.

Pesek and Kirshner considered teachers' reasons for not teaching conceptually. They had heard from some teachers that they do not have time; others had shared that conceptual teaching can take place only after students learn methods and rules—if there is time. The researchers conducted an experiment with six classes of fifth-grade students, who were grouped into two groups, but not by achievement. Both groups learned the area and perim-

eter of squares, rectangles, triangles, and parallelograms. One group received five lessons of traditional instruction, followed by three lessons of conceptual instruction; the other group received only three lessons of conceptual instruction. Both groups took a pre- and post-test and a delayed post-test. The lessons were observed, and students were interviewed.

In the traditional instruction, students were shown the formulas for finding the perimeter and area of squares, rectangles, triangles, and parallelograms. The teacher worked through questions; then students worked through more questions in groups. At the end of each lesson, the teacher reviewed the formulas.

In the conceptual teaching, which both groups received, students were asked to come up with their own ways to find area and perimeter, considering the relationships between the concepts. Students were invited to draw or measure with their hands, with square tiles, or with geoboards; they were learning the ideas with mathematical diversity.

The traditional instruction group received eight days of teaching, which many teachers think is ideal: five days of learning methods and rules, and three days of conceptual inquiry. The other group received only the conceptual teaching over three days. The results were stunning: the students who only learned for three days achieved at significantly higher levels on all the different assessments than those who learned for eight days.[32]

When the researchers investigated the reasons for this fascinating outcome, through interviews and test results, they found that students who learned traditionally had developed fixed ideas. For example, they associated the word *inside* with area and *outside* with perimeter. When they were asked which formula they would need to measure the amount of paint needed to paint a room, six students said they did not know or that they would need to know the perimeter because "walls don't have area; they go around." The researchers found that the students' learning of rules had interfered with their learning of concepts, probably because the mental effort to remember formulas and rules is great, and students appeared to focus on those rather than on the invitation to think, reason, and problem-solve.

This study seems particularly significant with regard to many teachers' belief that they do not have time to engage students deeply and conceptually. The results of this research demonstrate that it can take significantly less time, to engage students in this way (three-eighths of the time), and the time spent is more effective.

Of course, there are other problems with focusing on rules, as Jesús demonstrated above. When he resisted a conceptual explanation, asking instead, "What are the rules?" and "How do you know you are right if you don't know the rules?" he showed that when students think their role in mathematics is to remember and follow rules, they develop an unwillingness to engage in conceptual thought, some even thinking it is "not allowed."

Some teachers who watch my videos of students engaged in conceptual learning tell me that it is not possible for them to engage students similarly as their classes are too big. Given this reported barrier, I was thrilled when members of the engineering department at Stanford invited me to teach calculus to an incoming class of freshmen the summer before they started school. There were ninety-nine students in the class. Given the large class size, I invited members of my youcubed team to teach with me, a luxury I know that many teachers do not have.[33] Our goal for that summer was to teach the students with an approach of big ideas and connections, creating multiple opportunities for students to learn mental representations of calculus. Our job was made more challenging, and interesting, by the fact that many students had taken calculus to high levels in school, and some had not taken it at all. Student diversity in thinking and understanding is a challenge that all teachers face, no matter what grade level or content they teach or the extent to which schools have grouped students by prior achievement. Students are all different, I regard this difference as a wonderful resource in creating stimulating teaching and learning environments.

In chapter 3 I talked about Steve Strogatz, an applied mathematician at Cornell University, who has focused much of his life and work on sharing the beauty of mathematics with his students and with members of the public.[34] One of Steve's goals, similar to my own, is sharing mathematics as a subject of concepts and big ideas. Steve does this, and much more, in a book he has written about calculus, called *Infinite Powers*.[35] If you know anyone who is learning or teaching calculus, I recommend that you share this book, as it does something that many courses in calculus fail to do—it provides readers with a conceptual framing of the big ideas. This is what Steve has to say about big ideas in contrast to students' typical learning of calculus:

A single big, beautiful idea runs through the subject from beginning to end. Once we become aware of this idea, the structure of calculus falls into place as variations on a unifying theme. Alas most calculus courses bury the theme under an avalanche of formulas, procedures, and computational tricks. Come to think of it, I've never seen it spelled out anywhere even though it's part of calculus culture and every expert knows it implicitly. Let's call it the Infinity Principle. It will guide us on our journey.[36]

It is so interesting that Steve names the big idea of infinity, pointing out that all experts know it but also noting that he has *never* seen it highlighted in a calculus course. I am willing to bet that not only do all experts know this idea, but they also have mental representations that immediately come to mind when they think about the big ideas of calculus.

When the incoming Stanford students came to our calculus class that summer, it was clear that they had not been introduced to this idea, or to calculus conceptually, but to methods, rules, and procedures. My team and I decided we would teach students the big ideas of calculus, believing that this would be useful even for students who had taken calculus classes previously, as it would provide a conceptual frame into which they could place their knowledge of methods and rules and give them much-needed opportunities to develop mental representations and see mathematical connections. For those who had never taken calculus, we thought the approach would provide an accessible and meaningful introduction to the ideas. Steve's book was the course reader, and his chapters helped us design the class sessions focused on big ideas. Steve further describes the big idea of calculus in this way:

> Calculus proceeds in two phases: cutting and rebuilding. In mathematical terms, the cutting process always involves infinitely fine subtraction, which is used to quantify the differences between the parts. Accordingly, this half of the subject is called differential calculus. The re-

assembly process always involves infinite addition, which integrates the parts back into the original whole. This half of the subject is called integral calculus.[37]

To prompt students to think about the big idea of cutting into small parts and rebuilding into a whole—and to create mental models of this process—we adapted a lesson I had watched in Railside School. We invited students to come up with ideas for finding the volume of a lemon. One of the main applications of calculus is in finding the volume of complex, curved shapes.[38] In the students' prior learning of calculus, they may have sketched shapes, but this task gave students the opportunity to hold the complex shape in their hands and to infuse the ideas of calculus into their mental modeling of the shape. We grouped the students, gave each group a lemon, and invited them to investigate ways to find the volume of their lemon. The students were offered a range of investigative tools—a knife and cutting board, a vase with water, string, digital calipers, Play-Doh, protractors, and rulers.[39]

Figure 6.8 shows some of the students' ideas for finding the volume of a lemon.

That summer, as the students learned the big ideas of calculus, they were moved by the experiences of holding physical models in their hands; considering real objects, such as wheels and snowflakes; and applying the methods many of them had previously learned in school. One student, Sofia, reflected in an interview:

> The first problem that really helped open my eyes was the lemon problem. My group thought really creatively about the three methods that we tried, and actually physically

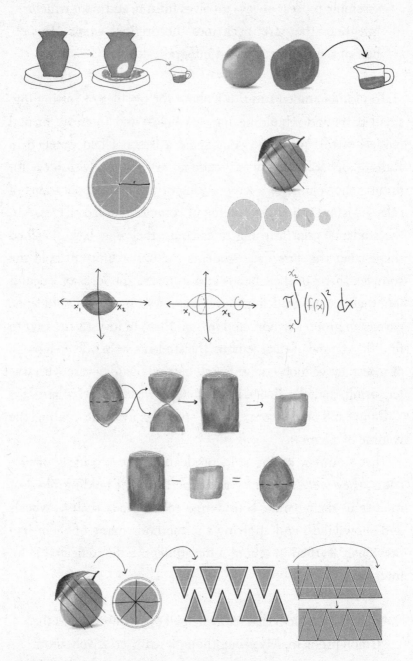

6.8 Different ways of visualizing summation and integration

manipulating the lemon helped me see why the different methods worked well. But it was at the end of it, when we discussed the problem as a class, that I saw all of my group's solutions were basically just different ways to perform summation and integration. It was the first time that I saw the integration formula and graph, and it actually made sense to me. Since that problem I have been riding a kind of high in the class. I now feel like if I try hard enough, and if I think creatively enough, then I can genuinely figure anything out.[40]

For this student, working with the lemon, and experiencing mathematical diversity, gave them the important idea that they could learn anything, which allowed them to ride "a kind of high" in the rest of the classes. Sofia's response also outlines the important teaching approach that I will turn to in the next chapter—the students investigated ideas, then a whole-class discussion was the location for methods to be introduced, connected, and reflected on.

Some of the students regarded the use of physical manipulatives, such as the lemons, balls, snowflakes, and minibikes we gave them to embody the ideas of calculus,[41] as "elementary" but they still reported that these experiences gave them a deeper understanding of the concepts they had previously only encountered as formulas to memorize.[42] Another student, Alec, told us that he had always been expected to just "accept the truth" of mathematical ideas; now his experience of "messing around with blocks and ropes and lemons like elementary schoolers" allowed him to genuinely understand ideas. This helped him develop conceptual understanding and personal ownership of the big ideas of calculus.

...

It is always exciting when I am invited to visit a school, but I received an invitation recently that was extra special. It came from Julie Shaw, the dynamic principal of a K–7 International Baccalaureate school on a First Nation Reserve in Canada. The Senpaq'cin School, which is open to any student, primarily serves the Nk'mip people, who are one of seven bands within the traditional lands of the Okanagan Nation, surrounding the Kelowna area in Canada.[43] I already had a sense from our email exchange that Julie was someone with a growth mindset, but I became convinced of this when the flight Cathy Williams and I were taking to the school was canceled. Julie sprang into action looking for other routes, airlines, and cars. Fortunately, between us, we worked something out and arrived at the school the next morning, excited and honored to be included in the activities of the day.

The morning started with a drumroll—a ceremony performed daily by different students, teachers, and community members and watched by the whole school. The chief of the band, Clarence Louie, chatted with me after we arrived and watched proudly as the students performed their drumming. It was a cold day in Kelowna, but no one seemed to mind as the students drummed and chanted together, giving a clear sense of unity.

Julie had asked Cathy and me to teach two lessons, one to students in grades three to five, and one to students in grades six and seven. I started the lessons by sharing the important messages we know from neuroscience: I told the students that their learning was limitless and that brain pathways were growing, strengthening, and connecting all the time. I told them that struggle and challenge were the most important times for them, that math was not about speedy thinking, and that all mathematics problems can be seen and solved in different ways. I also told them that approaching mathematics with diverse ways of seeing and solving problems, with multiple representations, results in important brain connections. We then worked on a dot card number talk (presented in chapter 5) to help that message come alive. The students listened intently to the brain messages and then readily shared the different ways they saw the seven dots.

Following that activity, Cathy and I shared that we had been enjoying mathematical beauty in several pieces of Indigenous art, which we showed them. Cathy then shared a picture of a dream catcher (fig. 6.9) and asked the students what the image meant to them. The students talked about the role of the dream catcher in their own lives and in their culture. Cathy then asked them to think mathematically about the image, applying a mathematical lens to the world, which is an important activity.

6.9 Dream catcher

The students in all classes were excited to share what they saw in the picture, including mathematical shapes such as triangles, trapezoids, and circles, as well as items from the world, such as a laptop, a river, and a room. When Cathy asked, "How many triangles do you see?" the students decided the answer was "an infinite number" as every triangle can be cut in half and then in half again. Some people might look at the shape and say there are no actual triangles, as the edges of the triangular shapes are curved. But we were happy to approach the shape with an ish lens (discussed in chapter 4). There are triangle-ish shapes on the dream catcher. If a mathematical lens is to be useful, we need to embrace the ish quality of numbers and shapes in the world, as well as the more precise versions. It is likely that the ish versions will be much more useful to people in their lives. Importantly, we should be helping students learn and use both the imprecise and the precise versions of shapes and numbers, giving them the ability to move between the big picture and focused thinking.[44]

Our visit to the school was magical, and I was honored to learn more about the culture of the Okanagan people and work with the students and teachers. But the collaboration became even more special when I received a follow-up email from Lisa van den Munckhof, the sixth- and seventh-grade teacher who had watched our lesson with her students that day. She shared the work she had undertaken following our visit. She first invited students to design their own dream catchers, giving them space to access their prior knowledge, use their own thinking, and activate what Zaretta Hammond calls their own "cultural capital" (fig. 6.10).[45] This work focused on the big idea of mathematical patterns.

After students had made their dream catchers and studied the patterns within them, Lisa artfully took their thinking into the realm of variables, helping them describe their patterns using

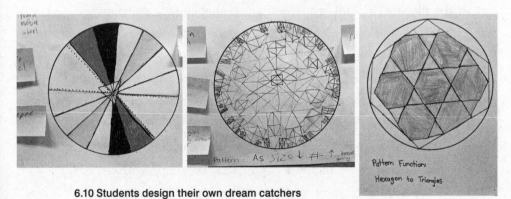

6.10 Students design their own dream catchers

algebra. Lisa shared that this work was difficult for the students, and they were "in the pit," which is the best place to be for learning and growth. She said that she had even made a mistake at one point in the discussion, which the class celebrated with high fives, as they had learned to do whenever students or teachers made a mistake. Importantly, Lisa shared that the activity had allowed "great connections" for the students. I am not surprised that the teachers and students were inspired by the connections they met between patterns and algebra, as it is inside mathematical connections that real beauty is found. The patterns the students created were their own, they were culturally meaningful, and they made the abstract concepts of algebra come alive.

Our experience of working with the school and learning about the value of preserving and honoring Indigenous people and culture was so powerful that it prompted a new initiative working with Indigenous educators across the world, to share many different examples of Indigenous art and mathematics.[46]

The approach I have set out in this chapter—of teaching and learning in ways that allow people to develop conceptual and connected understandings—is a key part of mathematical diversity. The rich tasks give students opportunities to struggle, allow them to develop mental representations, and help them access deep conceptual understanding that they will use for the rest of their lives.[47]

The approach of thinking conceptually and seeking connections is not just for school students; we can all take this approach to acquiring knowledge, which is likely to lead us to great beauty and insight. One of my favorite methods for developing conceptual insights into big ideas is journaling or making sketchnotes. Jotting words or visuals is a really good exercise for listening to ideas, connecting them, and keeping track of them.

In chapter 4 I shared that I often show a video of one of my former Stanford undergraduates, Yasmeena, solving a mathematics problem visually, in a way that highlights mathematical connections. What is interesting to me is the impact of the video—people are literally wowed by the beauty of the mathematical concepts and connections that they can see in visuals and movement. I am passionate about changing students'—and adults'—experiences of mathematics, as I know this beauty is available to us all when we approach mathematics as a set of concepts, big ideas, and connections.

7

DIVERSITY IN PRACTICE AND FEEDBACK

I hope that, by this point, you, your students, or your children have tried approaching mathematics with a lens of diversity and "ish"-ness, thinking conceptually about connections between ideas with deep, metacognitive thoughts and asking the very important question *Why?* as often as possible. I hope that you are now approaching mathematics in learning and in life with this flexible, conceptual approach. Maybe you have even had some good conversations with others, proposing conjectures and taking on the role of a skeptic. I hope that if you are a teacher or parent who is new to mathematical diversity, you will try all these ways of interacting with mathematics and perhaps start inviting students to experience mathematics in these diverse and engaging ways.

Over the years, many people have tried some of these ideas and come back to ask for further guidance. I am often asked questions such as these:

- Now that I have engaged students in these ways, with investigations and discussions, do I give them a worksheet to practice?

- After students have explored, do I give them a test to know if they have learned the ideas?
- How do I assess students well?

This chapter is going to answer these important questions, emphasizing that when students practice or work on assessments, they should continue to encounter mathematical diversity and engage deeply, strategically, and conceptually.

DIVERSE, DELIBERATE PRACTICE

One of the experts who has helped shape my thinking on productive forms of practice is Anders Ericsson,[1] a Swiss psychologist and professor at Florida State University. He and his coauthor, Robert Pool, famously described an important part of the process of becoming an expert in any area as engaging in "deliberate practice."[2] Ericsson defines this as practice that is purposeful and leads to specialized mental representations, with a clear feedback loop on ways to improve. When Ericsson released his work and claimed that practice was so important to the development of expertise, those who promote traditional teaching claimed that their approaches were ideally suited to enable this practice. But the opportunities that are provided in traditional math classrooms fall short in many ways. Deliberate practice is the engagement with *meaningful* ideas, through which students develop *representational models*, and it includes a clear *feedback loop* to provide opportunities for improvement.[3] In a traditional mathematics classroom, students practice content without meaning, diversity, or challenge; they are not encouraged to develop representational models; and the test feedback they receive is a blunt score, with no information on ways to improve. Fortunately, we can do much better, and when we do, students flourish.[4]

A CASE OF MATHEMATICAL DIVERSITY

When I applied to study for a PhD at King's College London, I knew what I wanted to investigate. I had spent the past two years studying for a master's degree in mathematics education, taking evening classes after teaching students in an urban London public secondary school during the day (grades 7–12). I felt ready for the challenge of a PhD and wrote a proposal to a highly competitive funding program, which, if I were successful, would provide a scholarship to fund my time studying mathematics teaching and learning. Some months later I learned that I had won the award and made the decision to leave my teaching job and become a full-time student again for the next few years.

The proposal that I had successfully outlined was my plan to study two different approaches to mathematics teaching and learning and to collect evidence on the effectiveness of each. I had noticed that there was a lot of debate and controversy over ways to teach mathematics, but very little supporting data or scientific evidence. I decided to try to help by contributing data and evidence to the field. I followed a cohort of students for the next three years, from the age of thirteen to sixteen, the end of compulsory schooling in the UK. The students were in two different schools, which I chose on the basis of their being very similar in terms of achievement and demographic data but totally different in terms of how they taught mathematics. I spent hundreds of hours in the classrooms of the two schools, collecting multiple forms of data. These included classroom observations, interviews with teachers and students, assessments of student understanding, and analyses of examination performance.[5] This detailed study of how different teaching approaches impact learning won the best PhD in education award in the UK from the British Educational Research Association (BERA).

In one of the schools (which I called Amber Hill), the teachers used a typical approach to mathematics teaching, in which the teacher explained methods to students and students then practiced them by working through textbook questions. The students received hundreds of hours of practice time, and their teachers were well qualified and supportive. In the other school (which I called Phoenix Park), teachers introduced ideas and activities to students; gave them the opportunity to investigate the ideas; invited them into a whole-class discussion, which developed and connected ideas; and then summarized the main ideas.[6]

As an example, the students at Phoenix Park were learning about the mathematical topic of a locus. In mathematics this is defined as a particular point, position, or place and usually extended to include all the points on a plane that are a certain distance away from a specific place. This is a mathematical tool that can help students explore relationships between points, lines, and curves. This becomes useful when people need to predict and analyze biological or social systems in scientific and other fields. Students at Phoenix Park began learning this topic by considering the locus of single points, then moving to an understanding that all the possible places in a locus from a point become a circle.

In typical US textbooks, the concept of a locus is introduced with a definition, and then students practice with narrow questions. At Phoenix Park, students were introduced to the concept

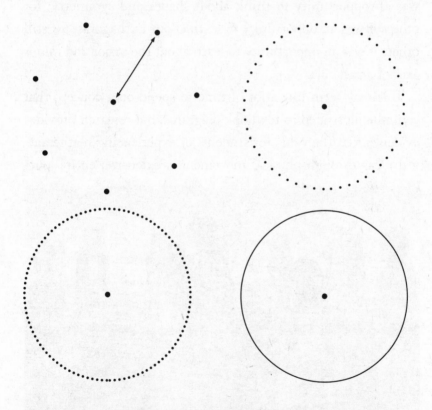

7.1 Example of how Phoenix Park students learn about the topic of a locus

of a locus by going into the playground and being asked to stand in different places. At first, they were asked to stand five meters away from the teacher, and the group of students saw that they formed an approximate circle.

The concept was extended, and students were asked to stand five meters away from a line. Then they were asked to stand at equal distances from two different points. They spent an entire lesson thinking about the concept of loci and experiencing the ideas physically through movement. Phoenix Park did not group the students by achievement, and the problems were always open enough for students to take in different directions, depending on their knowledge and understanding. For some students this task was an opportunity to think about shapes and symmetry; for others it was an opportunity to learn about Pythagoras; for still others it was an opportunity to learn about the major and minor axes of a parabola.

This may seem like a lot of time to spend on a concept that could be illustrated in textbook diagrams, but research provides evidence that the value for students of experiencing mathematical ideas through physical movement is extensive; entire jour-

nal volumes and books have been published about these strong mental representations, which we should want to instill for all mathematical concepts.[7] For example, one study observed students learning about negative numbers by folding paper (fig. 7.2):

7.2 Folding paper to highlight negative numbers

The researchers found that the experience of physically manipulating paper enabled students to develop mental representations of integers. This led students to achieve at higher levels on tests of not only negative numbers but also fractions and algebra.[8]

After the students at Phoenix Park experienced the concept of loci through standing in different positions in the playground, they were asked to practice the ideas for homework. Students were asked to consider the path of a dot (drawn on a piece of circular card) if it were rolled along the flat surface. They were asked to also think about a dot traveling on a triangle, a square, and a shape of their own choosing. They were asked to vary the position of the dot on the shapes and consider the paths formed.

This activity was important for the students at Phoenix Park for many reasons (e.g., its ish-ness and mathematical diversity) but I would like to focus on the diverse nature of the practice. The homework that students were given did not consist of repeatedly working on similar questions; it asked them to apply their understanding to different shapes. This not only meant that the practice was applied; it also meant that the work was challenging, giving students opportunities to struggle. The practice had other critical qualities: when students drew loci in relation to a triangle and a square, they were considering what researchers have termed "contrasting cases."

SEEING MORE

Sarah Levine and Dan Schwartz, colleagues of mine in Stanford's Graduate School of Education, both share the importance of contrasting cases for the development of expertise. For example, you might ask people to describe the object in figure 7.3:

Most people will say it is a pair of scissors.

But if you ask them to describe the two objects in figure 7.4, they are likely to tell you that the scissors on the left are short scissors, with a plastic handle, that are

7.3 A pair of scissors

not very sharp, have blunted ends, and are possibly children's scissors. They might describe the scissors on the right as long, sharp scissors with a hook on one of the metal handles.[9] The details of their observations only come about because contrasting cases are presented.

7.4 Two different pairs of scissors

This principle can be applied to many situations in life, including, of course, mathematics.

If you ask students to describe the shape in figure 7.5, they will likely say it is a triangle.

But if you ask them to describe the shapes in figure 7.6 . . .

7.5 A triangle

7.6 Two different triangles

. . . they are likely to tell you that the shape on the left is an equilateral triangle, or a triangle with three approximately equal sides and equal angles, and the shape on the right is an isosceles triangle, or a triangle with two equal sides and two equal angles, with a different orientation from the triangle on the left. Of course, the important part of considering contrasting cases is not just the stating of words but also the thinking that goes with the words. Researchers have found that the consideration of contrasting cases increases students' understanding significantly.[10]

Another example is illustrated in the two scenarios below, providing contrasting cases in two areas of mathematics that students often find difficult: percentages and fractions. Importantly, the questions are about the conceptual idea rather than calculations.

Show students the two scenarios and ask them to answer providing reasoning for their choice. If they are working together, which is ideal, they can reason with each other and then present their mathematical proofs, which then become further opportunities for discussion and learning.

7.7 Which of the two girls gets the bigger allowance? Construct a proof, with reasons. Include pictures, words, and numbers.

7.8 Which of the girls wants more of the cookie? Construct a proof, with reasons. Include pictures, words, and numbers.

In mathematics learning we should give students contrasting cases to consider as often as possible. Students can think about why they are similar and different, highlight features of the different mathematical ideas, learn from the differences they see, and provide reasons for their choices, all of which are valuable experiences that help create deep learning and expertise.[11]

I recently visited a friend in San Diego. I wanted to know what it was like living there, and I asked him a question that I realized later came from my belief in contrasting cases. I asked him if he had ever lived anywhere else. When he told me he had lived in Boston and in Santa Barbara, I was satisfied and asked him what it was like living in San Diego. I realized later that I was asking him about other places because without them, how would he know which features of San Diego are interesting or noteworthy?

At Phoenix Park, the students were often asked to consider contrasting cases. They practiced the idea of a locus by tracing the path of different dots on moving triangles, squares, and shapes of their own. This request to consider a locus in relation to different moving shapes will have encouraged them to see and understand how a locus interacts with properties of different shapes. Students learned more than if they had only considered a locus in relation to a circle, which is more typical of the mathematics practice that students are given.

The Phoenix Park teachers chose to give students an experience of practice that was focused on the big idea, the idea of a locus, not on small methods, which helped the students develop visual mental representations, building on the physical representations they experienced in class. The practice included an opportunity for students to create their own shapes, which brings agency into the mathematics learning process. When students have an opportunity to choose directions in their work, they are using their human agency, which brings about deeper engagement.[12] The simple act of inviting students to create their own shapes evokes this important learning quality.

The different qualities of the practice in which those students engaged contributed to their success in examinations—and life.[13] They thought about ideas through investigations and projects, teachers introduced new methods as they became relevant to the students, and students practiced their new understandings by applying the ideas to new and different situations. Whether students were learning new content, investigating ideas, or practicing with ideas, they were actively engaged through mathematical diversity.

PROCEDURAL VERSUS CONCEPTUAL QUESTIONS

You may not be surprised to learn that the students at Phoenix Park achieved at significantly higher levels on applied problem-solving assessments than the students who learned traditionally. What may surprise you is the fact that they also achieved at significantly higher levels in the traditional national examinations—a series of short questions students answer under timed conditions.[14] In England, national exams are very important; they are administered by examination boards to students at age sixteen. As part of my doctoral work, I was able to investigate my two sets of students' achievement by studying their answers to the national examination questions. I was helped by my PhD adviser, Paul Black, a very important person in the UK, who managed to get me permission to sit in a small, windowless room in the examination offices and read all the students' submitted papers. (The students had already received their results.) In analyzing which questions the students answered correctly or not, I saw something fascinating.

Before spending my day in that tiny room, I had divided all

the questions into two categories. If a question could be answered by the simple reproduction of a method, I labeled it a procedural question. If the question required something beyond the reproduction of a method, such as adapting a method, reasoning about the situation, or problem-solving, I labeled it a conceptual question. For example: "Calculate the mean of this set of numbers" was classified as a procedural question, as students did not have to choose or adapt a method; they just had to remember how to calculate a mean. In contrast, a conceptual question in the exam was as follows: "A shape is made up of 4 rectangles, [and] it has an area of 220 cm². Write in terms of x the area of one of the rectangles."

Figure 7.9 shows students' results for these two types of questions.

The students at Amber Hill achieved well on the procedural questions but poorly on the conceptual questions, which are usually more difficult. The students at Phoenix Park achieved at the same level on both sets of questions—and at significantly higher levels on the conceptual questions than the Amber Hill

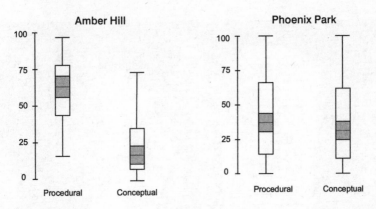

7.9 Students' results on examination questions that were procedural or conceptual

students. This higher level of work not only gave them overall higher examination results but also had a significant impact on their futures in work, in further study, and in their lives. It is also notable that the inequities that were present for the students at Phoenix Park were eliminated by that school's approach, but the inequities present for the students at Amber Hill, the traditional school, were reproduced in the national exams.

Interviewing students at Phoenix Park made clear that they achieved at higher levels in the exam because they had learned in school to use and apply methods and to think about their meaning. This was highlighted for me in an exam question that required students to use simultaneous equations (which are called systems of equations in the US). At Amber Hill, the students had repeatedly practiced a method for simultaneous equations. When faced with the exam question, they tried to use the method, but most of them jumbled the procedure and got the answer wrong. The Phoenix Park students were successful on that question even though they had not been taught a formal method, as they approached the problem with what I now describe as a growth mindset: they worked out solutions using and applying other methods they had learned.

I asked Angus, a student in year 11 (grade 10 in the US) at Phoenix Park, whether he felt there were topics and questions in the exam that he had not met in class:

> Well sometimes I suppose they put it in a way which throws you, but if there's stuff I actually haven't done before, I'll try and make as much sense of it as I can. Try and understand it and answer it as best as I can, and if it's wrong, it's wrong.[15]

Years later, I conducted a follow-up study with the students, who were now young adults of about twenty-four years of age. This study demonstrated the continuing effect that the Phoenix Park mathematics approach had: it enabled the students to be more successful as they applied their knowledge and positive mindsets to their jobs.[16] A categorization of their jobs revealed that the Phoenix Park students rose to significantly higher levels on the socioeconomic scale (SES). The graph in figure 7.10 shows the jobs the young adults held when I interviewed them, compared to the jobs their parents had at the time of the initial study. The difference in upward mobility is significant.

In their interviews, the former Phoenix Park students related their successes in life (specifically, job seeking and work) to the flexible approaches they had learned in their mathematics classes and to the responsibility they had been given to work problems out. As they chatted about work demands, they shared that they had been asked to take responsibility in different jobs, which they could do because they had learned to take responsibility in

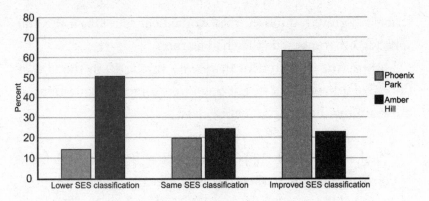

7.10 Analysis of students' jobs, compared to their parents'

their math classes. They told me that the flexible mathematics approach had also helped them know that if they were not feeling fulfilled in a job, they should seek a different one. Often, we teachers think we are teaching students mathematics so that they build good mathematics knowledge, but we are always doing more: teaching students an approach to life. The research I conducted with these young adults showed that the approaches and messages they had experienced in the classrooms of Phoenix Park had helped them in later years.

These students learned something that is important for all of us to remember: if we approach new challenges by thinking that we can have a go, apply something we have learned, and, as Angus shared, "make as much sense of it" as we can, then we will be more successful in our learning and in our lives.

Giyoo Hatano and Yoko Oura are professors in Japan who have added a great deal to the world's understanding of expertise. They describe two types: People who have developed "routine expertise" are able to solve familiar problems quickly and accurately but fail to go beyond what they call procedural efficiency. In contrast, those who develop "adaptive expertise" can be characterized by their flexible, innovative, and creative competencies within the domain, rather than in terms of speed, accuracy, and automaticity of solving familiar problems.[17]

It was clear to me, and to Hatano and Oura, that the students at Phoenix Park had developed adaptive expertise, which in turn enabled them to be successful both in their national examinations and in their lives.

When students engage with visual and physical representations, as the students learning about loci through movement did, they practice deliberately, and when they learn to adapt and apply methods in different situations, they develop adaptive expertise. This expertise turned out to be important for the young people attending Phoenix Park, many of whom had grown up in poverty but advanced to more financially stable situations, having learned in school the flexible approach to mathematics knowledge, which in turn taught them how to handle responsibility.[18]

APPLY DIVERSITY TO MATH EXAMPLES

Mathematical diversity can be brought into practice in another way. In traditional mathematics books, students are often shown many near-identical images and examples, but it is often more helpful to show them why an example does not work than to give more examples that do work. For example, when we teach students about birds, we often show similar types of birds, such as a sparrow, a hummingbird, and a dove. But it is more helpful to have students look at some flying creatures that are not birds, such as bats, as they learn about birds. The same principle applies in mathematics.

In my workshops for teachers, whenever we work on math together, I love nothing more than experiencing their excitement when they encounter diverse versions of the mathematics they teach. During the many discussions we have had over the years, they often describe the shape in figure 7.11 as an upside-down triangle. Sometimes I have cheekily responded, "You mean the shape that is also known as a triangle?" I am not surprised that teachers call this an upside-down triangle, as triangles are almost always shown with the narrowest point at the top, especially when they are introduced to students. Examples in books typically lack this diversity, which causes problems for learners.

For example, when a class of eight-year-olds was shown the picture in figure 7.12, they did not think it was a picture of parallel lines.

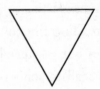

7.11 An "upside-down" triangle

And when they were asked if *a* is parallel to *c* in figure 7.13, most eleven-year-olds said, "No, because *b* is in the way."

Parallel lines are typically presented as shown in figure 7.14, which explains the students' reasoning:

Just as we should offer practice questions that apply methods

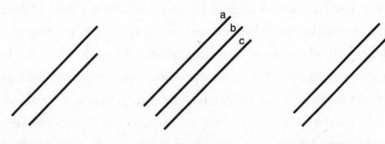

7.12 Parallel lines   7.13 Parallel lines   7.14 Parallel lines

and offer contrasting cases, we should also work to provide more examples of nontypical ideas and representations.

These types of practice appear to achieve the "deliberate practice" that Anders Ericsson describes as being meaningful and leading to representational models. To help others practice deliberately and effectively, the practice should involve as many of these qualities as possible:

Characteristics of Effective Practice

1. Application of methods: Problems should ask people to use methods in new and different situations.
2. Consideration of contrasting cases
3. Focus on concepts and big ideas, not small methods
4. Development of representational models that include visual or physical referents
5. Nonstandard examples and representations
6. Connections that people can see and learn; connections between mathematical ideas and between mathematics and the world

ASSESS WITH FEEDBACK LOOPS

Ericsson describes deliberate practice in terms of three qualities: the practice of *meaningful* ideas, the development of *representational* models, and a clear *feedback loop* to provide opportunities for improvement. The remainder of this chapter will consider the ways that students—and all people—can give and receive feedback productively, as I rarely see this in my visits to classrooms and businesses. Students and people in the workplace are told plenty of times that they are right or wrong, but this is not what is meant by a feedback loop.

Many teachers and parents lament the fact that their students or children do not engage metacognitively, using the different strategies I set out in chapter 2. Instead, students want to find an answer immediately, or they give up. This response comes about when assessments only focus on answers. If we want to encourage a range of positive mathematical behaviors, our assessments should honor and reward these behaviors. This means we should set out productive mathematical behaviors and give students feedback (and a summative score if needed) on their use of different behaviors. Mathematical targets serve as learning signposts, which guide students along their learning journey and encourage metacognitive thoughts and understanding.

Nancy Qushair is the head of middle school math at an International Baccalaureate school in California. Nancy not only values mathematical diversity in her teaching; she also provides opportunities for students to learn from their own mathematical behaviors by giving them meaningful feedback on their learning.

Nancy, like many educators, noticed that students emerged from the global pandemic with lower mathematics confidence and weaker problem-solving abilities.[19] Her response was not to double down on mathematics rules and procedures or to lecture students more—as some less-experienced teachers did—but to engage students in the mathematical modeling cycle shown in figure 7.15, which gives students signposts that turn their assessments into an iterative process of learning and adapting. I became so curious about Nancy's model of teaching and assessing that I accepted her invitation to visit her school and meet with the students myself.

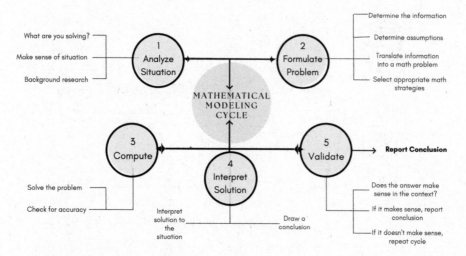

7.15 Mathematical modeling cycle

The unit of work Nancy developed was centered on a question about water conservation, an important topic for teenagers in California, which Nancy learned about in a teaching article.[20] Nancy had already established a mistake-and-struggle-friendly culture in her classroom, showing her students many of our videos that share the value of struggle.[21] With this important foundation, she posed a question: *Which uses more water, a shower or a bath?*

She told the students that people disagree on this issue, and it was their job to investigate and collect data to make their case. The students were invited to find out about showerhead flow rates and the different sizes of bathtubs in different homes, and choose whichever rates they wanted—giving them flexibility and choice, as well as many opportunities for ish numbers to be considered and valued. The new ideas Nancy taught the students included core algebraic concepts of linearity and generalization; importantly, she taught the students these concepts as they were grappling with data and investigating the rates they had chosen—evoking the impor-

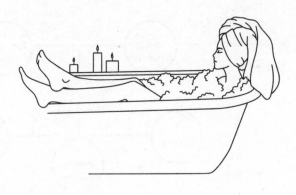

7.16 A shower or a bath?

tant teaching principle of teaching ideas *during* the time students work on tasks. This allows students to develop mental representations of the rates and the algebraic ideas through encountering them in a meaningful, real-world context. They learned connections between ideas such as constants and variables, and they focused on the big idea of generalization. But what happened in the final part of the model was most interesting to me.

I visited the school on a day when students were sharing with parents one of their most meaningful projects from the school year. Many of the students had chosen their math project—their conservation of water task. When I sat down with the students after their presentations, I was thrilled to hear them talk about various parts of the project that were important to them. Many talked about the value of considering something real from the world. They talked about how they had learned that mathematics was all about finding patterns, which is such an important perspective for students to develop. They also shared how much they had valued working together in groups. But what really struck me was the students' excited reports of the ways they had been invited into their own learning journeys—to be metacognitive (see chapter 2). Nancy achieved this important feat with a number of teaching moves.

First, she gave the students the representation of the learning process—the modeling cycle shown in figure 7.15—to guide their work and their learning. Many of the students said that seeing this model helped them in the learning process. As Ben described,

> I think it's pretty helpful because [the model] breaks it down. Usually, it's just "solve the problem" but it breaks it down into five main categories. It's "analyze the situation, use background information, actually compute, and then analyze your answer, and then report your conclusion." And then within that, it breaks it down into smaller parts of it so it's just easier to follow.

A number of the students reported that the model not only had given them a guide through their problem-solving process, but it had encouraged them to learn more deeply. Nota described this, saying,

> We looked at the problem and made sure that we were trying to solve not just what it was asking for but what it was deeply asking for.

Taylor, another of the thoughtful teenagers I sat with that day, noted that the math modeling cycle was not something that applied only to mathematics, sharing that it could be used "for basically anything," highlighting the generative nature of the modeling cycle.

Second, Nancy provided a rubric for assessing the students' work, which achieved something invaluable. Students used these descriptions as signposts, helping them see learning as an iterative process of working, revising, and improving. Taylor described this well:

> I mean, I think with the cycle too, it's not really a step-by-step thing. Like if you mess up or you make a mistake somewhere, you can go back and then add that part or use it to help you. I don't really think it's just step by step. I mean, you could use it that way if you want to, but it can also be used to go back and reevaluate your work, I think.

Third, the students in Nancy's classroom received feedback on their work regularly, but all the feedback had the quality of being oriented to their continued learning. Nancy wrote comments on students' work, something I always regard as a great gift—a teacher's insights into ways students can improve. Braden spoke about his appreciation of this feedback:

> I also love the teacher's comments because I always sometimes make mistakes. I know everyone here has made a mistake before and I feel that just looking back at your mistakes and the comments just helps me make it better and do better next time.

Braden captures the nature of the iterative process of learning he had come to know. To him and to all the teenagers I met that day, learning was a journey that was helped by the signposts and maps Nancy gave them to guide them along their way. They were aware of their own learning journeys and what they needed to do to improve. Ben contrasted this informed, iterative learning journey with the mathematics education he had known in his previous school:

> Before I came here, we did the project, and you would just receive a grade. It would be like you got an eight or whatever your grade was. There wasn't usually teacher comments, there wasn't any feedback whether it was good or bad. It was always a teacher's grade.

The students in Nancy's class enjoyed their task of working out the best ways to conserve water, they learned algebraic concepts through a real-world task with data, and they benefited from working in groups and being pattern seekers. But what may have been most meaningful was the fact that students came to view their learning as an iterative process of work, revision, and improvement. They had been given the tools to help them know where they were in their learning journey, and they had learned to be metacognitive.

The qualities Nancy built into her teaching and assessing—giving students feedback on their work and giving them opportunities to revise and improve their work—have been found to be some of the most important practices for encouraging growth mindsets in students.[22]

Conrad Wolfram, a friend and colleague from the UK, and his brother Stephen Wolfram, and their respective teams, have contributed a great deal to the world of applied mathematics. For example, they created the mathematics software Mathematica[23] and the mathematics tool WolframAlpha,[24] which not only helps anyone learning or working on mathematics; it also powers Siri, Alexa, and ChatGPT. In addition to this incredible work,

Conrad has offered a great deal to mathematics education; for example, he gave a compelling TED Talk in which he shared a modeling cycle as a guide for learning all mathematics, which is similar to the one Nancy's students used (fig. 7.17).[25]

Conrad points out that in mathematical work in the world, people need to learn how to interpret situations and define a question; they then need to turn the question into a form that is computable, perform the computation, and interpret the results. He points out that in classrooms, students focus only on the third part of the cycle—the computation. But with the advent of widely available technologies, that part of the process is arguably the least important to focus on. Conrad not only makes these arguments; he and his team have created an entire high school mathematics approach that expects students to calculate with technology, rather than by hand, and uses the time that becomes available to engage students in diverse mathematics problems, teaching them how to

1. DEFINE QUESTIONS

2. ABSTRACT TO COMPUTABLE FORM

3. COMPUTE ANSWERS

4. INTERPRET RESULTS

7.17 Mathematical modeling
Conrad Wolfram

set up the problems and use tools to calculate, then interpret, and analyze the results.[26] Examples of engaging problems are designing drones and investigating bias and fraudulent behavior.

The modeling process Conrad and his team invite students to engage with is one that any young person can use in their future schoolwork and employment. Nancy taught her middle school students in a similar way, providing them with opportunities to not only learn the modeling process but also monitor their own work and receive feedback along the way. Although this process is incredibly valuable for students to learn, it is absent from most high school mathematics courses.

TEACH WITH FEEDBACK LOOPS

One of the most important qualities of a feedback loop is that it is not focused on personal performance: that is, it does not communicate whether a person is right or wrong. Instead, it is focused on the work that is being learned or presented—and whether the idea needs revision. This change in approach is underscored by Elizabeth Bjork and Robert Bjork, two cognitive scientists who highlight the importance of frequent self-testing.[27] They point out that the act of retrieving information from our brains is one that makes information more readily available in future situations. But they emphasize that the productive form of testing that creates "desirable difficulties" is nonevaluative, so they suggest self- or peer-testing. Importantly, the testing and feedback loops that are successful are those that move the focus off people and performance and onto ideas.

I heard one of the most interesting cases of students being pro-

vided with a feedback loop that achieved this quality from a colleague in Stanford's Graduate School of Education. Carl Wieman, a Nobel Prize–winning physicist, was working at the University of Colorado, Boulder, when he became interested in education.[28] He had noticed that "bright, successful graduate students" were often clueless about physics until they spent some time working in labs, developing hands-on experience; then they started to become experts. Wanting to make sense of this, he turned to science and found that research from neuroscience, cognitive science, and education points to a better way to teach and learn than the lecture approach his students had experienced. This started Wieman's work in education, with the mission of bringing more active learning to college students' experiences of science.[29] He now holds a joint appointment in physics and education at Stanford.[30]

A few years earlier, when Wieman was at the University of British Columbia, he and his team conducted a fascinating experiment. They compared a typical lecture approach with a teaching approach designed around the principles of deliberate practice. The teaching was done in a first-year physics course, in the second term. The lecture approach was delivered by an experienced lecturer. The deliberate practice approach was taught by an inexperienced postdoctoral student. In the deliberate practice approach, students were given clickers, which they used to answer questions. The students read short articles before class. In class they were organized into small groups, where they discussed the ideas and answered questions about the material with the clickers. The questions targeted concepts that students typically found difficult. The teacher then displayed the results and talked with students about them. Sometimes the teacher would ask students to discuss the concept a second time, having shared some new ideas.

The two sections of the course were almost identical in terms of students' prior achievement and experience. At the end of these units, the average achievement of the students who received a lecture was 41 percent; the average achievement of the students who experienced deliberate practice was 74 percent. The researchers asked the students what they thought about the approach they experienced, knowing that students in college often resist approaches to teaching that are not lecture-based. Responses showed that 90 percent of students enjoyed the new approach, and 77 percent of that group said they would have learned more if the whole physics course has used that approach.[31]

It is helpful in all learning to get feedback on ideas, but the quality of the feedback is critically important. In the case above, students were invited to vote on ideas as a group, then they were shown correct and incorrect answers, which became the focus of a discussion.[32] This kind of feedback loop is perfect, as it is not evaluating students; the focus is the concept being taught. Student ideas can be gathered with technology, such as Google Forms or Quizlets, or without technology, such as voting on paper. The important thing is that the students evaluated different options by working together and voting on ideas as a group.

The feedback loop described in this example is not typical—it is very different from students being told they are right or wrong by a teacher. Another important feature of Wieman's teaching example is that the researchers focused on teaching the concepts that students typically find difficult. Teachers and parents who want to design questions for students—to accompany bigger projects, for discussion purposes, or for assessment—can learn much from this research, as well as from the research on contrasting cases and students applying their knowledge. All of the studies suggest that even mathematical practice and feedback need to employ the concept of mathematical diversity.

This chapter has focused on the importance of meaningful deliberate practice and feedback. I have shared a few different teaching examples that illustrate these qualities, including at Phoenix Park, in middle school classes in California, and in the college classes Wieman and his colleagues studied. The cases are different, but they each illustrate something important: that even practice and assessment, important parts of the learning experience, can and should be experienced with mathematical diversity. When they are, they provide important opportunities for students to learn concepts, through applied problems with mental representations, and to receive feedback as they learn, which provides guidance and opportunities to improve. In the final chapter, I will pull together the different ideas I have communicated so far and share a model of teaching that achieves them all, along with some inspiring stories of those who are teaching mathematical diversity in powerful ways.

8

A NEW MATHEMATICAL FUTURE

The examples I have set out in this book so far share different, important qualities of effective teaching, learning, and assessing. I started by sharing the importance of students learning how to learn with empowering mathematics strategies and the importance of becoming comfortable with times of struggle. I talked about the value of students learning both precise and "ish" numbers and shapes, and encountering multiple opportunities to develop representational models. I shared the importance of learning mathematics as a set of big ideas and connections, approaching numbers and shapes with flexibility, and, finally, I have shared that when students practice ideas, and get feedback, the ideas and feedback should be diverse and involve applications of mathematics. These different ideas come from many different sources, and one of the goals I have held in writing this book is to combine the advice on education that comes from different fields—education itself, but also psychology, cognitive science, and neuroscience. These different ideas combine into a model of teaching and assessing, represented in figure 8.1.

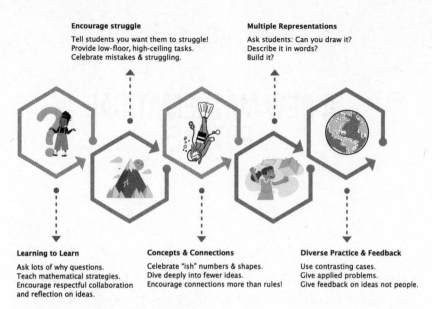

Encourage struggle
Tell students you want them to struggle!
Provide low-floor, high-ceiling tasks.
Celebrate mistakes & struggling.

Multiple Representations
Ask students: Can you draw it?
Describe it in words?
Build it?

Learning to Learn
Ask lots of why questions.
Teach mathematical strategies.
Encourage respectful collaboration
and reflection on ideas.

Concepts & Connections
Celebrate "ish" numbers & shapes.
Dive deeply into fewer ideas.
Encourage connections more than rules!

Diverse Practice & Feedback
Use contrasting cases.
Give applied problems.
Give feedback on ideas not people.

8.1 Teaching for equity and expertise

A NEW MODEL FOR EQUITY AND EXPERTISE

This model is intentionally abstract so that it may be applied to different teaching situations; I hope the various examples of teaching I have shared will add some of the details and color that help ideas come alive in different grade levels and circumstances. A good way to practice the model, as previous chapters have set out, is to change the typical order of instruction, so that students get to explore tasks before they are taught new ideas. How you combine these different components depends on your teaching situation. The model is a reminder that whatever teaching approach is followed, it should include as many of these different components as possible.

As the caption says, this model encourages expertise and equity. Mathematics education is a highly inequitable system; it is impossible to overlook the fact that few students advance to higher-level pathways, and the students who do go forward do not reflect the diverse nature of our society.[1] Black and Brown students are "held back," even when they have the same achievement as their white and Asian counterparts, something that was clearly demonstrated by a San Francisco legal group.[2] This is not acceptable, so a central goal of the 2023 California Mathematics Framework, for which I was one of the writers, was to highlight these inequities and suggest ways to address them.[3] The framework received significant pushback from a minority of people who spread misinformation about it. But when it came before the California State Board of Education in Sacramento, it received widespread support from the educators who were present as well as from all the county offices and equity organizations across the state. In July 2023, the board unanimously voted it into policy.

I have been fortunate in my career to study teachers who work to address inequities through their teaching of mathematics and who have been incredibly successful at reducing or eliminating completely inequities in race, gender, and social class in their classrooms.[4] These teachers all use the approach shown in figure 8.1 because it turns out that when we open mathematics to encourage diverse ways of engaging with and seeing the subject, many more students are successful. Narrow mathematics has done a spectacularly harmful job of narrowing the group of students who advance to high levels, pushing students out of mathematics and all the STEM courses that require mathematics, as chapter 1 detailed.[5] Fortunately, we can do better. Diversifying mathematics encourages inclusion of more diverse groups of students, which challenges even the most persistent of harsh inequalities.

This model of teaching is also used in high-achieving countries such as Japan. Swiss researcher Stéphane Clivaz and Japanese researcher Takeshi Miyakawa studied the details of two fascinating cases in Japan and Switzerland.[6] The researchers shared that lessons in Japan typically follow a similar structure, which they describe in this way:

- Introduction: a problem is introduced
- Research: students study and work to solve the problem in groups
- Sharing: students' ideas are shared and developed as a class
- Synthesis: the mathematical knowledge to be taught is summarized

This structure includes the principle that has been found to be so powerful—teaching methods *after* students have explored the ideas through tasks.[7] The step that the Japanese teachers refer to as "research" is a time of investigation, when students can use their own intuition and thoughts. Only later, after students have shared their thinking during a class discussion, do teachers introduce new methods. In the study, Clivaz and Miyakawa found that teachers in Japan spent longer engaging students in class discussion than Swiss teachers. In Japan they call classroom-discussion *neriage*—and they regard it as the most important part of the lesson.

This Japanese structure shares qualities with the Wieman model that I presented in the previous chapter, particularly the opportunity for students to consider and discuss ideas and then learn new methods through subsequent discussion.[8] This was the instructional flow at Phoenix Park, and this is the structure that we employ in youcubed summer camps, which are now taught by teachers across the US and the world, with students achieving impressive results: We start by sharing rich tasks with students; then, after they work and encounter a need for new knowledge, we introduce the knowledge as they are working in small groups or in a whole-class discussion.[9]

A few years ago, I received an email from Alexei Vernitski, a mathematics professor at the University of Essex in the UK who had read one of my books.[10] He described how he had become inspired to innovate his teaching and switched from narrow mathematics to tasks that invite mathematical diversity:

> I read *Mathematical Mindsets* . . . , and since then I have never taught a traditional lecture again. I enjoy the new way of teaching, and I like seeing how the students' faces light up when they work on "Boalerized" tasks instead of traditional maths problems.

Alexei went on to collaborate with a neuroscientist and a psychologist to investigate the difference between narrow and diverse mathematics tasks, studying students' brains using electroencephalograms (EEGs), looking for stimulation of the brain areas that are associated with motivation. This interdisciplinary collaboration produced fascinating results. First it found that students who were given standard math problems in tests reported less interest in continuing the test as they answered more questions. By contrast, students answering more diverse mathematical problems became more motivated as they worked.[11]

Additionally, the EEG testing found stronger patterns of activity associated with motivation and engagement in the brains of students who were working through the diverse mathematical problems—shifting activity to the left side of the prefrontal cortex. In prior studies, this pattern of "motivation-related" brain activity had been shown to decline as students worked through typically challenging problems, but it increased when students worked on diverse mathematics problems. Given this strong evidence, the researchers concluded that problems that encourage multiple ways of solving them, including using visuals, create positive learning experiences for students.[12]

Alexei thinks very carefully about mathematics tasks, regarding them as opportunities for students to see and learn important mathematical principles. He gives students interesting and challenging problems and invites them to discuss them in pairs and groups. He expects them to find the problems challenging, and instead of pre-teaching what the students need, he uses the important principle of waiting for students to need new knowledge before he introduces ideas to them. Alexei has noticed the changes in his students' engagement, enjoying moments when their faces light up as they work on mathematics in this way.

Likewise, a group of engineers in South Africa tested the ideas I share for open and diverse mathematics with problems used in college engineering programs. They found that all of the ideas are applicable to university mathematics and share examples of the ways they transformed engineering problems to make them more diverse.[13]

DIVERSE ENGAGEMENT THROUGH
DATA INVESTIGATIONS

It was a cold winter's day in Northern California when I received an email from mathematician Sol Garfunkel.[14] The email brightened my mood, as Sol is a colorful and interesting individual. A university mathematician who has dedicated his life and work to mathematics education, Sol has hosted PBS series and served for the past few decades as the director of an award-winning organization, the Consortium for Mathematics and Its Applications.[15] Sol and I started communicating through videoconferences, and he would often share with me his snowy winter surroundings as we chatted. I love snow, so this made the conversations even nicer. One of the things I learned about Sol is that he created, along with many engaging mathematics resources, an international mathematics competition in data modeling for high school and college students.[16] You may be thinking that a mathematics competition is not very interesting or important to your life, but let me see if I can interest you with some stunning data.

One well-known math competition used in schools across the US and in many other countries is the Math Olympiad.[17] I really like Math Olympiad questions, as they are often creative and interesting, but I do not like that the test questions are given under high-pressure, timed conditions, which are perfect for putting off women and other deep thinkers.[18] Every year a team of the most successful students in the US are sent to the International Math Olympiad. In the past thirteen years, the US has not sent a single female student, nor a single Black or Latine student.[19] Another university math competition that produces similarly horrible inequities is the Putnam exam. This is known as the "most prestigious"

college math competition, and it has a median score of zero out of 120 possible points.[20] The timed test consists of short, difficult questions. If you peruse webpages showing those who are successful in the Putnam, you will not see any women at all, nor will you see any racial diversity.[21] A young computer scientist at Stanford told me that when she was in her undergraduate years, students were required to state their Putnam score each time they walked into a math department meeting. This, to me, is a form of abuse—something meant to make people feel that their worth is judged by their performance on a narrow, stressful test.

In the midst of this bleak portrayal of mathematics tests that produce gendered and racialized results so severe that they should prompt alarm bells to ring throughout math departments is a shining ray of light. Sol has designed a competition that assesses mathematical modeling. Over a four-day period in each year of his competition, approximately 80,000 students work in groups of up to three people on difficult applied and diverse mathematical problems. Examples of problems include analyzing renewable energy in different states, examining trends in global languages, and planning an optical helicopter-search pattern. When students engage in these problems, they work with mathematical diversity and math-ish—drawing from different areas of mathematics, thinking in different ways, collaborating with each other, and building on each other's ideas. Impressively, 43 percent of participants are women, and 43 percent of winners are women.[22] At first the competition was designed for college students, but in the third year it was won by a team of high school students—the organizers had not even known that a high school had entered the competition. Ever since that time, this powerful mathematics experience has attracted and welcomed more and more high school teams.[23]

I became interested in this unusual math competition when Sol asked if my team at Stanford could investigate why the competition produces much more impressive gender results than all the other university math competitions. We set about to answer this question with a mixed-method study that included over forty-two hours of observations, interviews with faculty and students, and surveys of 1,327 students in ten countries. One of our findings was that students entered the competition because they felt they could bring their "full selves" and not be judged only by narrow mathematics.[24] The surveys and interviews were coded and analyzed to find the competition features that

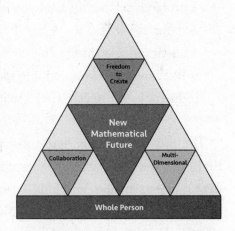

8.2 Students' reasons for entering the modeling competition

produced more equitable outcomes. This produced three themes, shown in figure 8.2, revealing that the students took part in the competition for three reasons—the opportunity to

- collaborate with others,
- engage in multidimensional mathematics and modeling, and
- create mathematical ideas.

One of the mathematics professors who enthusiastically recommends the competition to students at her college each year reflects:

> It is a different kind of experience compared to math competitions like the Putnam. In my opinion, it is a more accurate reflection of what professional and academic mathematicians do (reading, writing, working in a team, exchanging mathematical ideas, attacking problems that are not initially well-defined, spending time on a problem instead of having a shorter time period, etc.). Among other reasons, I recommend this competition for students who want to get a taste of what math "research" is like, and I recommend it to students who want to go to industry jobs directly after graduation.[25]

The professor's reflection and the results of our study both highlight the value of mathematical diversity, not only for student interest, success, and long-term learning but also for assessment. She shares an important point: this diverse experience is what real mathematics is.

The data competition—in addition to showing the ways that,

when we invite students to engage in more diverse content, a more diverse group of students are successful—shows the interest students have in data investigations. This is fortunate because we live in a data-filled world, as I share in this book, and any teacher of K–16 mathematics can diversify their content by infusing it with data. Temple Grandin, a professor of animal science and an autism advocate, makes a bold proposal to change the required content in high school and the early years of college from algebra to data analysis.[26] Research, including our study of Sol's data competition, suggests that such a change could diversify and increase student success and interest in STEM courses.[27]

THE IMPACT OF A SINGLE TEACHER

Over the years, I have advocated for a more diverse mathematical experience for students; I have met many parents and teachers concerned that they cannot make a difference when their children and students are experiencing so much narrow mathematics in school. They sometimes think they may as well go along with the system and teach in a narrow way too. I have two responses to this.

First, I know that when students are taught to see mathematics in diverse ways and learn to approach it from different perspectives, using the strategies I set out in chapter 2, they are changed from that time forward, and they reap great benefits, even when later classroom experiences are narrow. In chapter 4, I shared the story of Yasmeena, the undergraduate student who created a visual proof with Cuisenaire rods. When I asked Yasmeena if I could share her story in this book, she told me that the changed approach to mathematics she had learned in my class had enabled her to take and be successful in "multiple advanced math classes at Stanford (linear algebra, multivariable calculus, probability and stats)." Learning mathematics with mindset messages and diversity, whatever approach the teacher or parent chooses, sets students up for success in the rest of their lives.

My second response is that I know of countless K–12 teachers who have given students a diverse mathematical experience within the public school system, with its endless torrent of narrow curriculum standards and tests, and it made a huge difference to the students at that time and moving forward.[28] When teachers show students that they can see mathematics differently, it changes the ways students approach all future mathematics, narrow or diverse. One single teacher can make a huge difference for anyone, and I encourage you to be the person who makes that difference for the people you know—and yourself.[29]

It is probably not surprising that I am convinced of the impact of a single teacher, as I received this benefit in my own experience of learning mathematics. I attended a public "comprehensive" secondary school in England, and for most of my years in education, my mathematics experience was typical of many people's. I was successful and could compute methods at high speed, which was the way of working that was valued, but it did not interest me much. I was more interested in studying science at college, which meant that I chose mathematics as one of the "advanced level" or A-level subjects I studied when I was seventeen and eighteen. This is when I met Mrs. Marshall and mathematics changed for me.

Mrs. Marshall was quite a character, and she was considerably more personable than any previous math teacher I had ever experienced. She would often rush into our A-level classroom panting, having raced through the corridors to avoid the head teacher, as she was wearing dangly earrings, which the head teacher had banned, even for teachers. At the time, this act of rebellion impressed me, as did her chatty disposition. I remember noting to myself, with some surprise, that high-level mathematics could combine with an engaging personality! It was Mrs. Marshall's teaching that finally unlocked my mathematics potential and interest.

Mrs. Marshall used the same A-level mathematics textbook—filled with the ideas of calculus—as other teachers in the school, but she did not lecture on the methods and then ask students to compute with similar questions. She highlighted a few questions in each chapter and asked us to discuss them in groups. After we talked about the questions with our small groups, we would have a whole-class discussion. During this important time, Mrs. Marshall would interject new ideas and teach us new methods. This approach to mathematics learning—as a subject that could be seen differently, in which diverse student ideas and thoughts were valued, changed everything for me. It changed the way I

thought about myself, and it changed the way I thought about mathematics. This enabled me to consider mathematics as my future field of study and work.

What is interesting to me now, when I reflect on the experience I had as a seventeen- and eighteen-year-old student, is that the teacher transformed mathematics for me by changing two aspects of the mathematics learning experience: she invited students to talk about ideas, and she taught new methods after we had discussed situations and found a need for them, a practice that I have highlighted through several examples in this book.

DISRUPTING THE MATH STATUS QUO

Since that time, I have taught mathematics in London schools and in California and studied the teaching of mathematics in multiple settings in England and the US. My own teaching experience, and all the studies I have conducted, have shown the value of a diverse mathematical approach for students' learning and achievement.[30] But there are other benefits, just as important as achievement, that come from students' changed experiences with mathematics. I have termed one of these "relational equity" to capture a form of equality that comes about when students start to see mathematics as an opportunity for collaboration instead of competition, learn to treat each other with respect, and consider other viewpoints as they learn.[31] When we teach students to collaborate, carefully setting up group norms that teach them respect for each other, we make a huge contribution to the development of equitable societies. One of the goals of schools should be to produce young people who treat each other with respect; who value the contributions of others with whom they interact, irrespective of their race, class, gender, or any other difference; and who act with a sense of justice, considering the needs of others in society. A first step toward producing citizens who act in such ways is the creation of classrooms in which students learn to act in such ways.

In addition to this societal benefit, students learn to appreciate mathematical diversity and all that it can give them. Students from the different studies I have conducted describe the ways that their learning of mathematical diversity helps them achieve well. Seth was in one of the calculus classes I studied with Jim Greeno, a Stanford colleague, as part of an investigation into different approaches to calculus.[32] Students in Seth's class were invited to discuss ideas with each other; he reflects that this experience helped him later when he was working alone, as he had learned that if he was stuck, he should look at the problems in a different way. Simply working on problems with other students in class had given him an appreciation for mathematical diversity. The diversity Seth described to me provided a stark contrast to the reports from students in the other calculus classes in which they worked through narrow questions alone.[33] Yet this approach is so rarely a part of students' mathematics experiences, especially at higher levels.

Other students have helped me realize yet another value provided by mathematical diversity. Laquinita was a middle school student, age thirteen, who attended the first summer math camp I ever taught in the US. A school district had organized the camps for failing students, and attendance was compulsory. Laquinita's report card, which accompanied her, described her as "over exuberant." We found Laquinita to be thoughtful and engaged, and she was often willing to share her thinking, which we appreciated. At the end of summer camp, Laquinita contrasted her experience with her regular school math experience:

> It's like the way our schools did it is very black and white, and the way people do it here, it's like very colorful, very bright. You have very different varieties you're looking at. You can look at it one way, turn your head, and all of a sudden you see a whole different picture.[34]

Laquinita's description beautifully captures another value of mathematical diversity, beyond achievement and engagement. Mathematical diversity helps students develop mathematical appreciation, something that is given too little attention in the school system.[35] Some people prefer "black and white" mathematics, but they are vastly outnumbered by those who are inspired by the beauty of "colorful" and "bright" mathematics.

In one study of high school mathematics approaches, I met Toby, a seventeen-year-old senior at Greendale High School who was learning math through the Integrated Mathematics Program (IMP), an approach that teaches integrated mathematics, not separated algebra and geometry, through rich and complex mathematical situations.[36] When I observed classes, I saw students working together in multidimensional ways, building on each other's ideas, working together to find solutions. They were often passionately engaged, using high-level mathematical language as they debated different approaches to problems. When I took my Stanford colleague Jim Greeno, a world-famous cognitive psychologist, to watch one of the classes, he simply described it as "magical." At the end of the study, I sat down with

Toby and asked him to describe mathematics to me in his own words:

> Math is really beautiful and has these patterns in it that are amazing. Most art is just made up of patterns anyway. And so I've written a lot of poems about it, and a lot of songs involving it. Polyrhythms was one thing that kind of interspersed music and math for me—because it's like patterns that take multiple measures to repeat because they don't fit evenly over four bars, and it's exactly like a fraction because if you take a fraction high enough there's going to be common denominators. And so, seeing how patterns can be interesting and artistic. And math intersperses a lot for me that way.

Toby's mathematics approach impacted him in many ways. I saw in my classroom observations how mathematical diversity helped him develop understanding, and his description of mathematics in art and music gives us a deep sense of his mathematical appreciation. Toby also started to see the world with a mathematical lens; he describes mathematics as a set of ideas that weave through the world, interspersing music and art, providing patterns that he found "beautiful" and that gave meaning to his musical creations. The value of this life lens cannot be overstated. I find it baffling that there were people who worked to drive this approach out of schools, not seeing the value in the mathematical opportunities it provides. Fortunately, they were not successful.

All three of the students I quote talk about mathematics in atypical ways, as something colorful, even beautiful, and as a subject that is social, in which understanding is supported by the different ways people look at the ideas. They describe mathematics as a set of patterns that illuminate the world of art and music. Laquinita, at only thirteen years of age, captures mathematical diversity so well: "You can look at it one way, turn your head, and all of a sudden you see a whole different picture."

These conceptions of mathematics reflect its true discipline, yet they are, sadly, rare. Worse than this, our typical, narrow version of mathematics, which replays in classrooms nationwide, makes people who think differently believe that there is something wrong with them—that they are lesser.[37] But we can create something better for everyone—a mathematics that welcomes different ways of seeing and thinking and enables people to make mathematical connections and understand deeply. The students whose quotations I have shared all come from regular public school classrooms, and any one of us can help learners of mathematics—and ourselves—achieve something just as beautiful and as meaningful.

SYSTEMIC RACISM AND BIAS:
WORKING TO CHANGE THE STATUS QUO

Some readers know that my messages of mathematical diversity—particularly the idea that all students deserve access to high-level mathematics—have been met with considerable resistance, particularly from those who are successful in the current, inequitable system.[38] The traditional system of mathematics education sorts, ranks, and segregates students, and the narrow mathematics that is valued is easy to bring success for those who are wealthy, because they can pay for tutoring aimed at test success. Some people are very invested in keeping the system functioning in this way; they know that a mathematics approach that values and requires creativity and reasoning is less easy to coach, as it involves true understanding. Given this broader context, it may not be surprising that resistance to my ideas and my research evidence has taken the form of harassment, abuse, and, recently, even death threats to me and my children. But I have emerged from this tumultuous time stronger than I was before. This strength has come from the mindset I have developed, which has protected me through the harassment and abuse: I decided it is time to share the approach and strategies that I draw on. To conclude this book, I will share five principles that I believe are helpful to all of us, particularly those working to change inequitable systems.

I first encountered aggressive pushback and the spread of disinformation about my work after I published the results of the study showing that students at Railside School, a diverse, urban high school, achieved at higher levels than middle class students in a wealthier area who were taught traditionally.[39] The traditionalists working to block change claimed I manipulated data to produce this result, because it showed that when we change the way we teach and open pathways, many more students are successful. They went on to make the same claim about my study in England—a study that had won awards for its rigor.[40] When I was invited to be one of the writers of a new mathematics framework for the State of California, the spread of misinformation began again, and the pushback went up several notches to include death threats.[41] It was a Friday night when my email inbox started filling up with abuse. I quickly learned that Tucker Carlson had put my image on his show and ridiculed the fact that the proposed framework in California intended to promote social justice.[42] This was the beginning of a precarious few months, which included Stanford police adding my house to their daily patrols. Many people have expressed shock and dismay that a researcher working to produce evidence should experience this kind of abuse and harassment. Sadly, this kind of reaction is becoming more and more commonplace for academics;[43] scientists who study climate change receive similar harassment and abuse for their work.[44]

In addition to the threats, the group against the California Mathematics Framework worked to discredit my research studies, they tried to get journals to pull down my research papers, they spread misinformation about me in traditional and social media, they put misinformation on my Wikipedia page, and they convinced journalists to write multiple articles arguing against the framework and against me. I noted with great interest at that time that the rich and powerful in the US can direct and control the media.

I regard myself as a shy introvert. In my childhood I often

refused to speak with anyone outside my family and relied on my sister for all communication. As a young adult I avoided any public speaking and let others take the role as often as I could. Now, many years later, I speak to audiences of thousands, though never without nerves. What is remarkable to me, and something I would never have wished for myself, is the label I am given: "public figure." In recent years this label has even changed to a "controversial public figure." I never wanted to be a public figure and wanted even less to be in the middle of a public "war."

When I first saw myself described in this way in a news article, I reached out to the journalist and asked her to remove the word "controversial." She responded by saying that it was highly appropriate to describe me as controversial, as there are so many traditionalists who publicly disagree with my ideas, which contributes to the battles about math teaching. She highlighted other public figures whom she thought of as controversial, all people I admire greatly! Others have defined controversial people as those whose Wikipedia pages have received multiple edits.[45] My Wikipedia page has received not only multiple edits but edits so inaccurate and targeted that Wikipedia has locked the page to protect me.[46]

I have started to come to terms with the idea of being a "controversial public figure." But doing so—and even coming to see the value in embracing and enjoying the notoriety—has been a journey of change in my own mindset. Many people have asked me how I cope with attacks on my work, how I keep going in the face of death threats and abuse. My survival and my emergence as a stronger person, even more determined to fight for educational equality, is due to a set of ideas that I consider important for all, so I will end this book with them.

FIVE PRINCIPLES FOR BECOMING
AN EFFECTIVE CHANGE AGENT

1. Believe in Yourself

The first principle when working to change education, or other inequitable systems, concerns the importance of believing in yourself. The US, as well as other societies, has a culture of disrespecting educators, as well as people of color; women; queer, nonbinary, and trans people; those with physical disabilities; and anyone else who is different from "typical." If you are in more than one of these categories, the disrespect is magnified.

But educators, despite the lack of respect we get, have specialized knowledge that no other professionals have. Lee Shulman, a distinguished professor, introduced to the world a form of knowledge he termed pedagogical content knowledge, often shortened to PCK.[47] This knowledge, concerned with ways to teach well, lies at the intersection of content and pedagogy. For example, how do we share knowledge in a way that is most understandable to learners? What representations highlight the idea best? What misconceptions are typical? How do we deal with mistakes? Some university professors know their own content to very high levels but completely lack PCK, so they do not teach well. Good teachers have highly specialized PCK, which may take years to develop and is best developed inside the practice of teaching. Deborah Ball, a former dean of the University of Michigan and an education specialist, describes learning to teach outside of teaching as being analogous to learning to swim on a sidewalk.[48]

Some people think that teaching is not an intellectually challenging job. When they share this thought with me, I counter by posing a particular scenario of teaching as an example. Imagine that you are starting a class discussion on a mathematical idea with thirty students. Perhaps you ask the class a question, and a student offers an answer. In that moment you have many decisions to make as you form your response; you consider different questions: What did the student understand? How does that understanding connect with the mathematics I am discussing? How does the student's thinking connect with the broader mathematical horizon? Where could it lead? What does the student most need mathematically as a follow-up, but also, what do the other students in the class need? All of these thoughts must inform the teacher's next question or statement, and the decision must be made in a split second, while thirty people are watching them, waiting for them. Very few jobs require this complexity of thought and decision-making at high speed. Of course, this is just one small situation; teachers also need to know how to engage students deeply in all the content they are learning, which requires knowing their content up and down, inside and out, in ways that most people do not know content. I am glad that Lee Shulman identified pedagogical content knowledge and raised its status to an important place.

Despite the considerable expertise and knowledge that teachers draw from their practice and their years of study, many non-teachers believe they know better what should happen in the classroom because they went to school.[49] This is my advice to educators who face people who have very little information and are opposed to new, diverse ideas: *Know that you are the one with the knowledge and expertise, and believe in yourself.* Educate people who challenge you about the complexity of teaching, and explain some of its nuances. Don't be afraid to highlight your pedagogical content knowledge. Most educators shy away from flexing their expertise, but it may be time to share examples, including vignettes, that highlight the value of your educational decisions.

2. Practice Empathy

My second piece of advice is to practice deep, intentional empathy as much as you possibly can. An interdisciplinary group of Stanford professors and students conducted research on the ways we may promote healthy dialogue even when crossing political divides. After considering four studies involving 4,780 people, they found that when people communicate with empathy for their opponent's positions and ideas, they are much more likely to influence their thinking.[50]

Consider saying that you understand people's concerns and you have some examples that might be helpful for them to consider as they continue their own thought processes. Many of us value diversity in people and ideas, but we are not as welcoming as we should be to ideas different from our own.[51] If we truly value diversity, we should start conversations by welcoming other perspectives and talking them through. When I think about conversing with people whose views are opposite to mine, I remember the Buddhist advice that it is more helpful to be like a willow tree than a firm and strong tree. When it starts to snow, both trees have to bear additional weight on their branches. As the snow piles up, the branches of the firm tree stay rigid, until eventually they snap and break. The willow tree bends with the snow, accepting it, until eventually the branches spring back up, fresh and renewed.[52] I have written elsewhere about the importance of being flexible when approaching new situations and different viewpoints.[53] Sometimes flexibility helps us achieve the most difficult of goals.

At this point I will share that none of the people who opposed the California Mathematics Framework and worked to discredit me have been open to discussing ideas. If they were, I truly believe that we would have experienced much more agreement than disagreement, and their ideas would have made more of a difference. I always welcome respectful challenges; lively debate is a sign of a healthy and productive community, and that is how we all learn. What are not healthy are personal attacks and attempts to discredit not the ideas but the person.[54]

3. Build a Network

My third area of advice is about appreciating the incredible value of other people and communication. If you are working in a difficult area, I strongly suggest you find supportive allies; these can be any people in your life—friends, family, colleagues. I have found over the years that people who are attacked, including myself, have a natural tendency to turn inward and stay silent. This is unfortunate because the most restorative and generative approach when the pressure is on is to connect with others. When I eventually started to talk about the attacks on my work, I was contacted by hundreds of women scientists who shared similar experiences of harassment and defamation.[55] That support from people in similar situations changed everything for me.

4. Investigate

My fourth area of advice is to collect and share data—which can take many forms. One of the greatest changes I have seen came about when a Toronto school principal, who was committed to the value of the growth mindset principles permeating teaching, interviewed students, filming their answers to questions asking them how they felt about math. He played the videos for teachers, which led to widespread changes in teaching approaches.[56] If you are observing an aspect of your system that may need changing, collect data. As you do so, look at what you find through the lens of diversity. As the above example illustrates, data can take many different forms—student pathways, student achievement, racial inequities, students' feelings after taking timed, narrow math tests—all of which can be powerful in effecting positive outcomes.

5. Develop a Warrior Mindset

My fifth and final piece of advice may be the most important of all as it concerns our mindsets and ways of framing pushback. My learning about this practice draws from both Buddhist and Taoist teaching, though I am not an expert in either religion. Of course, both traditions reject the idea of being an expert, and they position even the wisest communicators of ideas as people who are themselves constantly learning. In what follows, I share my interpretation of an idea that helps me in my continued work to make education a fundamental human right for all students.

In both Buddhist and Taoist teachings, leaders frame the work of change as the work of a warrior. The root of the word *warrior* comes from *war*, but the Buddhist and Taoist conceptions of warriorship are not about fighting or war; they are about connecting with the world in new and different ways, oriented toward making a difference.[57] The awareness that comes from this framing of equity work can change the way we move through the world, empowering us to greater levels of effectiveness and protecting us from harmful forces who work to block change. Importantly, being a warrior is a state of internal awareness, a way of connecting differently with our own minds.

John Little communicates the philosophies of Bruce Lee, who was famous for his Hollywood appearances as a martial arts expert and was also noteworthy for his developed mindset and approach to life.[58] Little states that many people in the Western world neglect to connect with their warrior force, ignoring its presence, and are thus far less powerful than they could be. All who do connect with their inner warrior can often benefit from untapped inner powers that enable them to make connections and maximize their potential.[59]

Being a warrior involves a commitment to making change, not for your life or your children's, but for the world. To do this, you first need to acknowledge your own strength and goodness, in order that you may project that to others. Warriors are not endlessly positive or upbeat but have chosen to look outward from themselves and improve conditions, spreading good ideas and goodness in the world.

After my colleagues and I taught our first youcubed camp at Stanford, in which we shared the ideas of mindset, brain growth, and mathematical diversity, we later heard that the students had returned to their schools and shared the ideas with students who had not attended our camps. They told their math classmates that they should not give up but rather think that they had not learned something *yet*. They encouraged each other to think differently about math problems—to draw them or build them, for example. I have heard from other teachers about students who are passionate about sharing mindset ideas with others in their classes—these students are acting as mindset warriors: they are taking what they know is helpful information and turning outward to share it with others.

After acknowledging your value in what you can bring to others and committing to making a difference, you will need to connect with your authentic self. As we let go of fixed ideas and develop more flexible minds, we are more likely to become authentic presences in the world. Warriorship—the self-realization of your own power and potential to make good in the world—involves getting to know the real you, your honest self. It is more important to know yourself than it is to know anyone else. Self-knowledge allows you to interpret the world around you to its fullest. As you bring greater wakefulness into your mind, it will help you transcend doubt and hesitation about being your authentic self. I will not say more about this important way of being, other than this: authenticity is a state that, once attained, is never lost.

The next aspect of developing warriorship is centrally connected to the idea of yin and yang, a concept that is important in almost every ancient culture examined by modern archaeology, including Buddhist and Taoist religions.[60] Yin/yang captures natural dualities in the world, such as sun and shade, fire and water, or correctness and incorrectness; it conveys that these opposing ideas are interconnected in important ways. The opposites need to exist together, in balance. If you move too far along any continuum (toward yin, for example), you will be helped by experiencing some of the yang. Some of us grew up thinking we always need to be happy or strong or positive; we are not meant to be sad or weak or negative—but this mindset contradicts the natural balance of the world. No one can be endlessly happy, positive, or strong; it is important to realize this and to acknowledge the feelings that we may have tried to push away. Then, instead of pushing away feelings of negativity or helplessness, we acknowledge and feel them, so that we can return to a state of balance.[61]

Some people think that the idea of *warrior* means strength, but while strength is needed at times, it is not possible—or even desirable—to be strong all the time. Even warriors need to acknowledge their vulnerability. That feeling that we always need to be strong or successful, that we can never mess up or fail, is what causes people to give up their dreams and goals. Acknowledging the need for yin and yang in all aspects of warriorship, and life, can be extremely liberating.

Chögyam Trungpa, a Tibetan Buddhist monk and prolific writer, communicates the important ways warriors are in touch with the dualities of experience:

> The fullness of her experience is her own, and she must live with her own truth. Yet she is more and more in love with the world. That combination of love affair and loneliness is what enables the warrior to constantly reach out to help others.[62]

The concept of yin and yang has been helpful in reminding me that there is always a balance. As you work to spread ideas of change, you may receive a lot of positive feedback, but there will always be resistance too, and you should expect that—and even welcome it as a sign that your ideas have the potential to change something. People will not go to the trouble of pushing back unless they feel that your ideas will make a difference (which for some reason scares them). In my own work of sharing the value of a different approach to mathematics, I have frequently needed to draw on the warrior's courage when the going gets rough. The work is just too important to give up. A central part of being a warrior involves becoming more comfortable with different ways of being. I am strengthened by the knowledge that those who work to attack and discredit me lack understanding, not only of education and mathematics but of ways to live compassionately alongside fellow human beings whom they do not agree with. It is hard for me to be annoyed with someone who lacks understanding, as it signals that they have not had opportunities to learn and to develop. I also know that my strength and courage are balanced by my vulnerability, which I need to be comfortable with.

The concept of warriorship conveys complex ideas of mindset and outlook that give us different, mindful strategies for dealing with challenge. I communicated in chapter 3 the importance of walking at the edge of your understanding, because this is a place where the greatest knowledge development occurs. I see work in equity as walking on a different edge, the edge of change.

Those who have the potential to make change as they walk along this edge are often targeted. If you are walking on the edge of change, you will probably get arrows fired at you.

You must endure this to get to the other side, but usually the other side is a beautiful place to be. Don't let those arrows cause you to retreat or fall down. Just as it is important to be comfortable with the edge of struggle, it is important to be comfortable with the edge of change too. When people have not summoned up the mindset and courage required to walk on the edge, and the first arrows are fired, they climb back down to safety. This is part of the reason that important change does not happen.

Those working to make education systems more equitable are particularly vulnerable to attack because our school system is built on privilege. Many outdated practices that are still used in education hail back to older days before we had the evidence we now have of neuroplasticity, neurodiversity, mindset, and brain connectivity; these practices remain entrenched because they are supported by powerful people who benefit from them. If you can learn to welcome and reframe pushback as a positive sign, as an indication that you really can change something, you will have evoked the spirit of a warrior.

A few years ago, after I arrived in New Mexico ready to work with teachers on sharing the ideas of mindset and mathematical diversity, I looked out at the audience and immediately noticed two young women wearing T-shirts that held my attention.

I asked the teachers—Jana Ward and Zaira Falliner—what had prompted them to make these T-shirts with the slogan #trueBoaliever. They told me that they had just finished their master's degrees and were working as teacher leaders—sharing the ideas of mindset and mathematical diversity with local in-

structors. This was an exciting time, as there was a lot of enthusiasm from teachers and measurable results from students—both in their achievements and beliefs. Jana and Zaira had created a group called the math action team. But teachers who taught traditionally had started pushing back on the group, calling them a cult. I too have had this strange accusation leveled at me. Jana and Zaira's response was to lean into the accusation and make T-shirts declaring their true "Boalief"! Jana reflected that they both knew what was right for students, and they would not be deterred. This is a perfect example of the warrior spirit. Jana and Zaira were getting attacked and labeled—so they leaned into the ideas, accepted them, and made them their own.

Sensei Koshin, a Zen teacher, psychotherapist, and author, points out that in all great hero stories there is trouble. How people work with the trouble makes them who they are.[63] If you are an educator working to open access to all students, to raise up those students who have not been given opportunities, and to fight for the underprivileged in our society, you are one of these heroes, and your story will probably involve some trouble or challenge. Your hero story depends on your reaction to the trouble and challenge, especially the ways you use your knowledge and your mindset to turn those experiences into strengths.

I do not think my ideas are particularly controversial, even though others assign them this label. But if believing that all students can learn and that it is unacceptable for high-level mathematics classrooms in schools and colleges to be racially tracked means I am controversial, then I am willing to accept that label. In fact, I will proudly own it. If the work you are doing has the potential to disrupt the status quo, then you too should wear the label with pride. Because when you choose to reframe, accept, and own the labels given to you, you will do something important. Your warrior mindset will add a layer of imperviousness to the way you carry yourself through life. You will show the world that you are not going to be cowed, slowed, intimidated, or even bothered by what attackers have to say, because you understand where they come from and what motivates them. Jana and Zaira took this approach and were labeled a cult, so they leaned into the idea, proudly displaying that they were "Boalievers." This is the warrior spirit we all need to develop if we are working to promote equitable outcomes.

This book has shared qualities of teaching that encourage mathematical diversity and ishness and presented many different examples of how teachers achieved this in their classrooms. We started with the important goal of teaching students how to learn using some key metacognitive and mathematical strategies that any learner can use. We then considered the importance of embracing struggle and shared strategies for all of us to cultivate comfort with struggle. But our real journey into mathematical diversity began with identifying the most important areas of mathematics and the ways that each can be approached with multiple perspectives. I shared the value of ish numbers and shapes in learning and in life. We then considered the power of visual mathematics, with several examples spanning the grades. From there we explored mathematics as a conceptual and connected subject that is important to approach with flexibility. And we wrapped up with a discussion of diversity in mathematical practice, assessment, and feedback.

What I hope I have conveyed in these descriptions and cases is the fact that students are more interested and more successful when the content they are learning allows them to engage in different ways. This is important for the learning of all content, at all ages and levels. It is not the case that we have a nation of people who cannot be successful in math, but it is the case that we have many millions of people who would be much more successful and more engaged if they had experienced mathematical diversity and math-ish. Whether or not you need to adopt the spirit of a warrior to enact these ideas, strategies, and approaches, I hope they strengthen you as you go through life, allow you to see more and learn more in every situation you encounter, and enable you to lift others up to levels they did not even know they could reach, inspired by the beauty of math-ish and mathematical diversity.

ACKNOWLEDGMENTS

I would not be able to continue working on equitable mathematics approaches without my youcubed partner and best friend Cathy Williams. I owe Cathy an extraordinary debt (I will make it up to you, pal) for all of her work on the mathematics visuals in the book, for reading drafts of chapters, and for always being a sounding board for ideas. I could not have written this book without Cathy.

Jill Marsal, my literary agent, is always such a great help; thank you Jill for helping me believe I could write another book.

I am also grateful to the team at HarperOne, especially Maya and Shannon, who have both been so gracious in answering my numerous email questions and being endlessly positive about my ideas.

The book is also helped by Kane Lynch's ability to bring characters to life; thank you for all the beautiful illustrations, Kane.

Many teachers and other educators have shared ideas and images that I have included in the book and I am very grateful for their generosity and collegiality. I am fortunate to have known a few warriors in my life, who are endlessly working to make education a better and more equitable space, and many of them are inside this book.

Last but certainly not least, I am thankful for my amazing family, who let me escape to my writing retreat when needed. They even let me take our mini Bernedoodle, Dougal, who keeps me entertained with the endless puzzles she makes herself. My family helped me through some very hard times (that I talked about in chapter 8) with support, humor, food, FaceTime calls, and love. Oh, and the greatest advice always, including the wisdom I got from my youngest daughter, when the California Mathematics Framework passed, which was: "Dab on them haters, Mum!"

For additional information and resources,
visit www.mathish.org.

NOTES

Chapter 1: A New Mathematical Relationship

1. A. F. Cabrera, J. L. Crissman, E. M. Bernal, A. Nora, P. T. Terenzini, and E. T. Pascarella, "Collaborative Learning: Its Impact on College Students' Development and Diversity," *Journal of College Student Development* 43, no. 1 (2002): 20–34; H. Jazaieri, K. McGonigal, T. Jinpa, J. R. Doty, J. J. Gross, and P. R. Goldin, "A Randomized Controlled Trial of Compassion Cultivation Training: Effects on Mindfulness, Affect, and Emotion Regulation," *Motivation and Emotion* 38 (2014): 23–35; Organisation for Economic Co-operation and Development, *PISA 2015 Results*, vol. 5: *Collaborative Problem Solving* (Paris: PISA, OECD, 2015); M. F. Winters, *Inclusive Conversations: Fostering Equity, Empathy, and Belonging Across Differences* (Oakland, CA: Berrett-Koehler Publishers, 2020).

2. J. Boaler and M. Staples, "Creating Mathematical Futures Through an Equitable Teaching Approach: The Case of Railside School," *Teachers College Record* 110, no. 3 (2008): 608–45; R. K. Anderson, J. Boaler, and J. A. Dieckmann, "Achieving Elusive Teacher Change Through Challenging Myths About Learning: A Blended Approach," *Education Sciences* 8, no. 3 (2018): 98, https://www.mdpi.com/2227-7102/8/3/98; J. Boaler, J. A. Dieckmann, G. Pérez-Núñez, K. L. Sun, and C. Williams, "Changing Students' Minds and Achievement in Mathematics: The Impact of a Free Online Student Course," *Frontiers in Education* (2018): 26; J. Boaler, J. A. Dieckmann, T. LaMar, M. Leshin, M. Selbach-Allen, and G. Pérez-Núñez, "The Transformative Impact of a Mathematical Mindset Experience Taught at Scale," *Frontiers in Education* (2021): 512.

3. See Haverstock School, https://www.haverstock.camden.sch.uk.

4. M. Suárez-Pellicioni, M. I. Núñez-Peña, and A. Colomé, "Math Anxiety: A Review of Its Cognitive Consequences, Psychophysiological Correlates, and Brain Bases," *Cognitive, Affective, and Behavioral Neuroscience* 16 (2016): 3–22, https://pubmed.ncbi.nlm.nih.gov/26250692/.

5. C. Drew, "Why Science Majors Change Their Minds (It's Just So Darn Hard)," *New York Times*, November 4, 2011, https://www.nytimes.com/2011/11/06/education/edlife/why-science-majors-change-their-mind-its-just-so-darn-hard.html.

6. C. Edley Jr., "At Cal State, Algebra Is a Civil Rights Issue," EdSource, June 5, 2017, https://edsource.org/2017/at-cal-state-algebra-is-a-civil-rights-issue/582950.

7. J. Boaler, "Op-Ed: How Can We Make More Students Fall in Love with Math?" *Los Angeles Times*, March 14, 2022, https://www.latimes.com/opinion/story/2022-03-14/math-framework-california-low-achieving.

8. J. Boaler, *What's Math Got to Do with It? How Teachers and Parents Can Transform Mathematics Learning and Inspire Success* (New York: Penguin, 2015).

9. Anderson et al., "Achieving Elusive Teacher Change," 98; Boaler and Staples, "Creating Mathematical Futures."

10. See my bio at "Our Team," https://www.youcubed.org/our-team/.

11. Z. Clute, "Bad at Math No More," Hechinger Report, April 4, 2017, https://hechingerreport.org/opinion-bad-math-no/.

12. OECD, *Skills Matter: Additional Results from the Survey of Adult Skills* (Paris: OECD Publishing, 2019), https://doi.org/10.1787/1f029d8f-en.

13. OECD, *Skills Matter*.

14. L. Abrams, "Study: Math Skills at Age 7 Predict How Much Money You'll Make," *Atlantic*, May 9, 2013, https://www.theatlantic.com/health/archive/2013/05/study-math-skills-at-age-7-predict-how-much-money-youll-make/275690/.

15. J. Boaler, *Limitless Mind: Learn, Lead, and Live Without Barriers* (New York: HarperCollins, 2019).

16. "How to Learn Math for Teachers," Stanford Online, https://online.stanford.edu/courses/xeduc115n-how-learn-math-teachers.

17. J. Boaler, K. Dance, and E. Woodbury, *From Performance to Learning: Assessing to Encourage Growth Mindsets* (Stanford, CA: youcubed, 2018), https://www.youcubed.org/wp-content/uploads/2018/04/Assessent-paper-final-4.23.18.pdf.

18. E. K. Chestnut, R. F. Lei, S. J. Leslie, and A. Cimpian, "The Myth That Only Brilliant People Are Good at Math and Its Implications for Diversity," *Education Sciences* 8, no. 2 (2018): 65; S. Leslie, A. Cimpian, M. Meyer, and E. Freeland, "Expectations of Brilliance Underlie Gender Distributions Across Academic Disciplines," *Science* 347 (2015): 262–65.

19. M. Merzenich, *Soft-Wired: How the New Science of Brain Plasticity Can Change Your Life* (San Francisco: Parnassus, 2013), 2; N. Doidge, *The Brain That Changes Itself* (New York: Penguin, 2007).

20. T. Iuculano, M. Rosenberg-Lee, J. Richardson, C. Tenison, L. Fuchs, K. Supekar, and V. Menon, "Cognitive Tutoring Induces Widespread Neuroplasticity and Remediates Brain Function in Children with Mathematical Learning Disabilities," *Nature Communications* 6 (2015): 8453, https://doi.org/10.1038/ncomms9453.

21. L. Letchford, *Reversed: A Memoir* (Irvine, CA: Acorn, 2018).

22. J. Boaler, "Crossing the Line: When Academic Disagreement Becomes Harassment and Abuse," Stanford University, March 2023, https://joboaler.people.stanford.edu/.

23. 2023 Mathematics Framework, California Department of Education, updated October 20, 2023, https://www.cde.ca.gov/ci/ma/cf/.

24. Anderson et al., "Achieving Elusive Teacher Change," 98.

25. See, e.g., the work of Eugenia Cheng, Keith Devlin, Dan Finkel, Maryam Mirzakhani, Steve Strogatz, and Talithia Williams.

26. E. Cheng, "What If Nobody Is Bad at Maths?," *Guardian*, May 29, 2023, https://www.theguardian.com/books/2023/may/29/what-if-nobody-is-bad-at-maths.

27. L. Chen, S. R. Bae, C. Battista, S. Qin, T. Chen, T. M. Evans, and V. Menon, "Positive Attitude Toward Math Supports Early Academic Success: Behavioral Evidence and Neurocognitive Mechanisms," *Psychological Science* 29, no. 3 (2018): 390–402.

28. L. R. Aiken and R. M. Dreger, "The Effect of Attitudes on Performance in Mathematics," *Journal of Educational Psychology* 52, no. 1 (1961): 19–24, https://doi.org/10.1037/h0041309; L. R. Aiken, "Update on Attitudes and Other Affective Variables in Learning Mathematics," *Review of Educational Research* 46 (1976): 293–311, https://www.jstor.org/stable/1170042.

29. F. Pajares and M. D. Miller, "Role of Self-Efficacy and Self-Concept Beliefs in Mathematical Problem Solving: A Path Analysis," *Journal of Educational Psychology* 86 (1994): 193–203, https://doi.org/10.1037/0022-0663.86.2.193; K. Singh, M. Granville, and S. Dika, "Mathematics and Science Achievement: Effects of Motivation, Interest, and Academic Engagement," *Journal of Educational Research* 95 (2002): 323–32, https://doi.org/10.1080/00220670209 96607.

30. Researchers often use IQ as a measure, even though this test has racist origins. See, e.g., "History of the Race and Intelligence Controversy," Wikipedia, updated September 16, 2023, https://en.wikipedia.org/wiki/History_of_the _race_and_intelligence_controversy.

31. Chen et al., "Positive Attitude Toward Math."

32. S. Beilock, *How the Body Knows Its Mind: The Surprising Power of the Physical Environment to Influence How You Think and Feel* (New York: Simon and Schuster, 2015).

33. Chen et al., "Positive Attitude Toward Math."

34. C. B. Young, S. S. Wu, and V. Menon, "The Neurodevelopmental Basis of Math Anxiety," *Psychological Science* 23, no. 5 (2012): 492–501.

35. J. Boaler, "Prove It to Me!" *Mathematics Teaching in the Middle School* 24, no. 7 (2019): 422–28.

36. Boaler, "Prove It to Me!"

37. See "Our Team" at https://www.youcubed.org/our-team/.

38. Boaler et al., "Transformative Impact," 512.

39. Boaler et al., "Transformative Impact," 512.

40. Iuculano et al., "Cognitive Tutoring," 8453; Chen et al., "Positive Attitude Toward Math"; V. Menon, "Salience Network," in A. W. Toga, *Brain Mapping: An Encyclopedic Reference*, Academic Press, 2015, https://med.stanford .edu/content/dam/sm/scsnl/documents/Menon_Salience_Network_15.pdf.

41. C. S. Dweck, *Mindset: The New Psychology of Success* (New York: Random House, 2006); J. W. Stigler and J. Hiebert, *The Teaching Gap: Best Ideas from the World's Teachers for Improving Education in the Classroom* (New York: Simon and Schuster, 2009); H. Stevenson and J. W. Stigler, *Learning Gap: Why Our Schools Are Failing and What We Can Learn from Japanese and Chinese Education* (New York: Simon and Schuster, 1994); A. Ericsson and R. Pool, *Peak: Secrets from the New Science of Expertise* (New York: Random House, 2016).

42. Boaler et al., "Transformative Impact," 512.

43. P. Liljedahl, "Building Thinking Classrooms: Conditions for Problem-Solving," in *Posing and Solving Mathematical Problems: Advances and New Per-*

spectives, eds. P. Felmer, E. Pehkonen, and J. Kilpatrick, 361–86, (Switzerland: Springer, 2016).

Chapter 2: Learning to Learn

1. "John H. Flavell," Wikipedia, updated February 19, 2023, https://en.wikipedia.org/wiki/John_H._Flavell.
2. S. Moritz and P. H. Lysaker, "Metacognition—What Did James H. Flavell Really Say and the Implications for the Conceptualization and Design of Metacognitive Interventions," *Schizophrenia Research* 201 (2018): 20–26.
3. J. Boaler and P. Zoido, "Why Math Education in the US Doesn't Add Up," *Scientific American Mind* 27, no. 6 (2016): 18–19.
4. OECD Learning Compass for Mathematics, "The Future of Education and Skills: The Future We Want," https://www.oecd.org/education/2030/OECD-Learning-Compass-for-Mathematics-2023-13-Oct.pdf.
5. "Hattie Ranking: 252 Influences and Effect Sizes Related to Student Achievement," Visible Learning, n.d., https://visible-learning.org/hattie-ranking-influences-effect-sizes-learning-achievement/.
6. S. M. Fleming, "The Power of Reflection," *Scientific American Mind* 25, no. 5 (2014): 30–37.
7. E. Mitsea, A. Drigas, and P. Mantas, "Soft Skills and Metacognition as Inclusion Amplifiers in the 21st Century," *International Journal of Online and Biomedical Engineering (iJOE)* 17, no. 4 (2021): 121–32, https://doi.org/10.3991/ijoe.v17i04.20567.
8. A. Grant, "The Impact of Life Coaching on Goal Attainment, Metacognition and Mental Health," *Social Behavior and Personality: An International Journal* 31 (2003): 253–63, https://doi.org/10.2224/sbp.2003.31.3.253.
9. D. Wilson and M. Conyers, *Teaching Students to Drive Their Brains: Metacognitive Strategies, Activities and Lesson Ideas* (Alexandria, VA: ASCD, 2016).
10. P. Black and D. Wiliam, "Assessment for Learning," in *Assessing Educational Achievement*, ed. D. Nutall, 7–18 (London: Falmer, 1986).
11. C. A. Hecht, M. C. Murphy, C. S. Dweck, C. J. Bryan, K. H. Trzesniewski, F. N. Medrano, . . . , and D. S. Yeager, "Shifting the Mindset Culture to Address Global Educational Disparities," *npj Science of Learning* 8, no. 29 (2023), https://doi.org/10.1038/s41539-023-00181-y.
12. A. Vrugt and F. J. Oort, "Metacognition, Achievement Goals, Study Strategies and Academic Achievement: Pathways to Achievement," *Metacognition and Learning* 3 (2008): 123–46, https://doi.org/10.1007/s11409-008-9022-4; G. Özsoy, "An Investigation of the Relationship Between Metacognition and Mathematics Achievement," *Asia Pacific Education Review* 12 (2011): 227–35, https://doi.org/10.1007/s12564-010-9129-6; M. V. Veenman, R. D. Hesselink, S. Sleeuwaegen, S. I. Liem, and M. G. Van Haaren, "Assessing Developmental Differences in Metacognitive Skills with Computer Logfiles: Gender by Age Interactions," *Psihologijske teme* 23, no. 1 (2014): 99–113; Wilson and Conyers, *Teaching Students to Drive Their Brains*.
13. Wilson and Conyers, *Teaching Students to Drive Their Brains*.
14. J. Boaler, "Promoting 'Relational Equity' and High Mathematics Achievement Through an Innovative Mixed-Ability Approach," *British Educational Research*

Journal 34, no. 2 (2008): 167–94; Boaler and Staples, "Creating Mathematical Futures."

15. Boaler, "Promoting 'Relational Equity.'"
16. Boaler and Staples, "Creating Mathematical Futures."
17. Boaler, "Promoting 'Relational Equity.'"
18. S. F. Reardon, E. Weathers, E. Fahle, H. Jang, and D. Kalogrides, *Is Separate Still Unequal? New Evidence on School Segregation and Racial Academic Achievement Gaps* (Stanford, CA: Stanford CEPA, 2019); S. F. Reardon, E. Fahle, H. Jang, and E. Weathers, "Why School Desegregation Still Matters (a Lot)," *Educational Leadership* 80, no. 4 (2023): 38–44.
19. P. Cobb, T. Wood, E. Yackel, and B. McNeal, "Characteristics of Classroom Mathematics Traditions: An Interactional Analysis," *American Educational Research Journal* 29, no. 3 (1992): 573–604.
20. Wilson and Conyers, *Teaching Students to Drive Their Brains*; "Hattie Ranking: 252 Influences," Visible Learning.
21. M. Amalric and S. Dehaene, "Origins of the Brain Networks for Advanced Mathematics in Expert Mathematicians," *Proceedings of the National Academy of Sciences* 113, no. 18 (2016): 4909–17.
22. J. Boaler, "Paying the Price for 'Sugar and Spice': Shifting the Analytical Lens in Equity Research," *Mathematical Thinking and Learning* 4, nos. 2–3 (2002): 127–44.
23. E. Gray and D. O. Tall, "Duality, Ambiguity, and Flexibility: A 'Proceptual' View of Simple Arithmetic," *Journal for Research in Mathematics Education* 25, no. 2 (1994): 116–40.
24. Boaler, *Limitless Mind*.
25. Vrugt and Oort, "Metacognition, Achievement Goals"; Özsoy, "Investigation of the Relationship"; Veenman et al., "Assessing Developmental Differences"; Wilson and Conyers, *Teaching Students to Drive Their Brains*.
26. J. Boaler, *Mathematical Mindsets: Unleashing Students' Potential Through Creative Math, Inspiring Messages and Innovative Teaching* (Hoboken, NJ: Wiley, 2015).
27. Boaler, *Mathematical Mindsets*, 47.
28. T. LaMar, M. Leshin, and J. Boaler, "The Derailing Impact of Content Standards—An Equity Focused District Held Back by Narrow Mathematics," *International Journal of Educational Research Open* 1 (2020): 100015; Boaler, "Promoting 'Relational Equity'"; Boaler and Staples, "Creating Mathematical Futures."
29. Boaler et al., "Transformative Impact," 512.
30. Boaler, "Promoting 'Relational Equity'"; Boaler and Staples, "Creating Mathematical Futures."
31. E. G. Cohen, R. A. Lotan, B. A. Scarloss, and A. R. Arellano, "Complex Instruction: Equity in Cooperative Learning Classrooms," *Theory into Practice* 38, no. 2 (1999): 80–86.
32. Cohen et al., "Complex Instruction."
33. E. G. Cohen and R. A. Lotan, *Designing Groupwork: Strategies for the Heterogeneous Classroom*, 3rd ed. (New York: Teachers College Press, 2014).
34. Boaler et al., *From Performance to Learning*.

35. Boaler, *Mathematical Mindsets*, 141–70.
36. "An Example of a Growth Mindset K–8 School," youcubed.org, n.d., https://www.youcubed.org/resources/an-example-of-a-growth-mindset-k-8-school/.
37. Boaler, *Limitless Mind*.
38. My *Limitless Mind* videobook is available at LIT, https://litvideobooks.com/limitless-mind.
39. Boaler et al., *From Performance to Learning*.

Chapter 3: Valuing Struggle

1. C. S. Dweck and D. S. Yeager, "Mindsets: A View from Two Eras," *Perspectives on Psychological Science* 14, no. 3 (2019), 481–96; L. S. Blackwell, K. H. Trzesniewski, and C. S. Dweck, "Implicit Theories of Intelligence Predict Achievement Across an Adolescent Transition: A Longitudinal Study and an Intervention," *Child Development* 78, no. 1 (2007): 246–63; O. H. Zahrt and A. J. Crum, "Perceived Physical Activity and Mortality: Evidence from Three Nationally Representative US Samples," *Health Psychology* 36, no. 11 (2017): 1017; D. S. Yeager, K. H. Trzesniewski, and C. S. Dweck, "An Implicit Theories of Personality Intervention Reduces Adolescent Aggression in Response to Victimization and Exclusion," *Child Development* 84, no. 3 (2013): 970–88; J. A. Okonofua, J. P. Goyer, C. A. Lindsay, J. Haugabrook, and G. M. Walton, "A Scalable Empathic-Mindset Intervention Reduces Group Disparities in School Suspensions," *Science Advances* 8, no. 12 (2022): eabj0691.
2. J. A. Mangels, B. Butterfield, J. Lamb, C. Good, and C. S. Dweck, "Why Do Beliefs About Intelligence Influence Learning Success?: A Social Cognitive Neuroscience Model," *Social Cognitive and Affective Neuroscience* 1, no. 2 (2006): 75–86; J. S. Moser, H. S. Schroder, C. Heeter, T. P. Moran, and Y. H. Lee, "Mind Your Errors: Evidence for a Neural Mechanism Linking Growth Mind-Set to Adaptive Posterror Adjustments," *Psychological Science* 22, no. 12 (2011): 1484–89.
3. H. S. Schroder, T. P. Moran, M. B. Donnellan, and J. S. Moser, "Mindset Induction Effects on Cognitive Control: A Neurobehavioral Investigation," *Biological Psychology* 103 (2014): 27–37.
4. Dweck and Yeager, "Mindsets."
5. J. W. Stigler and J. Hiebert, "Understanding and Improving Classroom Mathematics Instruction: An Overview of the TIMSS Video Study," *Phi Delta Kappan* 79, no. 1 (1997): 14; Stevenson and Stigler, *Learning Gap*; Stigler and Hiebert, *Teaching Gap*.
6. Stigler and Hiebert, *Teaching Gap*.
7. S. Olson, *Countdown: Six Kids Vie for Glory at the World's Toughest Math Competition* (Boston: Houghton Mifflin, 2004), 48–49, https://steveolson.com/assets/countdown.pdf.
8. Merzenich, *Soft-Wired*, 2; N. Doidge, *The Brain That Changes Itself* (New York: Penguin, 2007).
9. Dweck and Yeager, "Mindsets."
10. C. S. Dweck, "The Secret to Raising Smart Kids," *Scientific American*, January 1, 2015, https://www.scientificamerican.com/article/the-secret-to-raising-smart-kids1/.

11. Hecht et al., "Shifting the Mindset Culture"; D. S. Yeager, J. M. Carroll, J. Buontempo, A. Cimpian, S. Woody, R. Crosnoe, . . . , and C. S. Dweck, "Teacher Mindsets Help Explain Where a Growth-Mindset Intervention Does and Doesn't Work," *Psychological Science* 33, no. 1 (2022): 18–32, https://doi.org/10.1177/09567976211028984; Okonofua et al., "Scalable Empathic-Mindset Intervention"; Dweck and Yeager, "Mindsets"; D. S. Yeager, P. Hanselman, G. M. Walton, J. S. Murray, R. Crosnoe, C. Muller, . . . , and C. S. Dweck, "A National Experiment Reveals Where a Growth Mindset Improves Achievement," *Nature* 573, no. 7774 (2019): 364–69; Blackwell et al., "Implicit Theories of Intelligence."

12. C. Good, C. S. Dweck, and J. Aronson, "Social Identity, Stereotype Threat, and Self-Theories," in *Contesting Stereotypes and Creating Identities: Social Categories, Social Identities, and Educational Participation*, ed. A. J. Fuligni, 115–135 (New York: Russell Sage Foundation, 2007); S. R. Levy and C. S. Dweck, "The Impact of Children's Static Versus Dynamic Conceptions of People on Stereotype Formation," *Child Development* 70, no. 5 (1999): 1163–80.

13. Blackwell et al., "Implicit Theories of Intelligence."

14. Zahrt and Crum, "Perceived Physical Activity and Mortality," 1017.

15. Yeager et al., "Implicit Theories of Personality Intervention."

16. Okonofua et al., "Scalable Empathic-Mindset Intervention."

17. Yeager et al., "Teacher Mindsets Help Explain"; Anderson et al., "Achieving Elusive Teacher Change," 98; P. Bui, N. Pongsakdi, J. McMullen, E. Lehtinen, and M. M. Hannula-Sormunen, "A Systematic Review of Mindset Interventions in Mathematics Classrooms: What Works and What Does Not?" *Educational Research Review* 40 (August 2023): 100554.

18. D. Coyle, *The Talent Code: Unlocking the Secret of Skill in Maths, Art, Music, Sport and Just about Everything Else* (New York: Random House, 2009).

19. Coyle, *The Talent Code.*

20. Boaler, "Prove It to Me!"

21. "Steven Strogatz," Wikipedia, updated October 12, 2023, https://en.wikipedia.org/wiki/Steven_Strogatz.

22. D. J. Watts and S. H. Strogatz, "Collective Dynamics of 'Small-World' Networks," *Nature* 393, no. 6684 (1998): 440–42, https://www.nature.com/articles/30918.

23. S. D. Levitt, "Steven Strogatz Thinks You Don't Know What Math Is," *People I (Mostly) Admire*, podcast episode 96, produced by Morgan Levey, Freakonomics Radio, January 6, 2023, https://freakonomics.com/podcast/steven-strogatz-thinks-you-dont-know-what-math-is/.

24. L. Deslauriers, L. S. McCarty, K. Miller, K. Callaghan, and G. Kestin, Measuring Actual Learning versus Feeling of Learning in Response to being Actively Engaged in the Classroom, *Proceedings of the National Academy of Sciences* (2019): 201821936. M. Kapur, "Productive Failure in Learning Math," *Cognitive Science* 38, no. 5 (2014): 1008–22; D. L. Schwartz, C. C. Chase, M. A. Oppezzo, and D. B. Chin, "Practicing Versus Inventing with Contrasting Cases: The Effects of Telling First on Learning and Transfer," *Journal of Educational Psychology* 103, no. 4 (2011): 759; D. Schwartz and J. Bransford, "A Time for Telling," *Cognition and Instruction* 16, no. 4 (1998): 475–522.

25. Deslauriers et al., "Measuring Actual Learning"; Kapur, "Productive Failure in Learning Math"; Schwartz et al., "Practicing Versus Inventing," 759; Schwartz and Bransford, "Time for Telling."
26. Deslauriers et al., "Measuring Actual Learning."
27. Dweck and Yeager, "Mindsets"; Deslauriers et al., "Measuring Actual Learning"; P. Barrouillet, "Theories of Cognitive Development: From Piaget to Today," *Developmental Review* 38 (2015): 1–12; Kapur, "Productive Failure in Learning Math"; K. Shabani, M. Khatib, and S. Ebadi, "Vygotsky's Zone of Proximal Development: Instructional Implications and Teachers' Professional Development," *English Language Teaching* 3, no. 4 (2010): 237–48.
28. Ericsson and Pool, *Peak*.
29. "Ken Robinson (educationalist)," Wikipedia, August 17, 2023, https://en.wikipedia.org/wiki/Ken_Robinson_(educationalist); Ken Robinson, "Do Schools Kill Creativity?," James Clear, transcript from TED Talk delivered February 2006, https://jamesclear.com/great-speeches/do-schools-kill-creativity-by-ken-robinson.
30. Merzenich, *Soft-Wired*, 2; Doidge, *Brain That Changes Itself.*
31. Coyle, *Talent Code.*
32. "Tasks," youcubed, n.d., https://www.youcubed.org/tasks/; "K–8 Curriculum," youcubed, n.d., https://www.youcubed.org/resource/k-8-curriculum/.
33. "Unlock Your Child's Limitless Potential with Math Education Based in Neuroscience," Struggly, n.d., https://www.struggly.com/.
34. See The Learning Pit, https://www.learningpit.org/.
35. E. A. Gunderson, S. J. Gripshover, C. Romero, C. S. Dweck, S. Goldin-Meadow, and S. C. Levine, "Parent Praise to 1- to 3-Year-Olds Predicts Children's Motivational Frameworks 5 Years Later," *Child Development* 84, no. 5 (2013): 1526–41.
36. C. S. Dweck, "Secret to Raising Smart Kids."
37. "Rethinking Giftedness Film," youcubed, n.d., https://www.youcubed.org/rethinking-giftedness-film/.
38. Hecht et al., "Shifting the Mindset Culture"; J. Feldman, *Grading for Equity: What It Is, Why It Matters, and How It Can Transform Schools and Classrooms* (Thousand Oaks, CA: Corwin Press, 2018).
39. Boaler, *Limitless Mind.*
40. S. Singh, *Fermat's Enigma: The Epic Quest to Solve the World's Greatest Mathematical Problem* (New York: Anchor, 2017), 6.
41. P. Brown, "How Math's Most Famous Proof Nearly Broke," *Nautilus*, May 21, 2015, https://nautil.us/how-maths-most-famous-proof-nearly-broke-235447/.
42. "How to Learn Math for Teachers," Stanford Online.
43. The correct answer is $11/12$. I would get this by converting $2/3$ to $8/12$ and $1/4$ to $3/12$.
44. "The Importance of Struggle," youcubed.org, n.d., https://www.youcubed.org/resources/the-importance-of-struggle/; "Excerpt of Jo from 'The Importance of Struggle,'" youcubed.org, n.d., https://www.youcubed.org/resources/excerpt-of-jo-from-the-importance-of-struggle/.
45. R. Kehoe, "A Secret of Science: Mistakes Boost Understanding," Science News

Explores, September 10, 2020, https://www.snexplores.org/article/secret -science-mistakes-boost-understanding.

46. Barrouillet, "Theories of Cognitive Development."
47. Shabani et al., "Vygotsky's Zone of Proximal Development."

Chapter 4: Mathematics in the World

1. Cabrera et al., "Collaborative Learning"; Jazaieri et al., "Randomized Controlled Trial"; OECD, *PISA 2015 Results*, vol. 5; Winters, *Inclusive Conversations*; Boaler and Staples, "Creating Mathematical Futures"; Anderson et al., "Achieving Elusive Teacher Change," 98; Boaler et al., "Changing Students' Minds and Achievement," 26; Boaler et al., "Transformative Impact," 512.
2. Boaler et al., "Transformative Impact," 512.
3. Reardon et al., *Is Separate Still Unequal?*; Reardon et al., "Why School Desegregation Still Matters."
4. Reardon et al., "Why School Desegregation Still Matters."
5. Boaler, *Limitless Mind*.
6. Boaler, "Promoting 'Relational Equity.'"
7. S. D. Levitt and S. J. Dubner, *Freakonomics: A Rogue Economist Explores the Hidden Side of Everything*, rev. ed. (New York: William Morrow, 2010).
8. S. Levitt and S. Dubner, "America's Math Curriculum Doesn't Add Up," podcast episode 391, produced by Zack Lapinski, Freakonomics Radio, October 2, 2019, https://freakonomics.com/podcast/math-curriculum-doesnt-add -up-ep-391.
9. Levitt and Dubner, "America's Math Curriculum."
10. S. J. Ball, "Education, Majorism and 'the Curriculum of the Dead,'" *Curriculum Studies* 1, no. 2 (1993): 195–214.
11. C. Everett, *Numbers and the Making of Us: Counting and the Course of Human Cultures* (Cambridge, MA: Harvard Univ. Press, 2017).
12. Everett, *Numbers and the Making of Us*.
13. Everett, *Numbers and the Making of Us*.
14. "Cuisenaire Rods: Gattegno and Other Films," Association of Teachers of Mathematics,n.d.,https://www.atm.org.uk/Cuisenaire-Rods---Gattegno-and -other-films#:~:text=Cuisenaire%20rods%20were%20invented%20in ,music%20with%20an%20instrument%20gave.
15. "Factorization Diagrams," Math Less Traveled, n.d., https://mathlesstraveled .com/factorization/.
16. W. H. Cockcroft, *Mathematics Counts* (London: HM Stationery Office, 1982).
17. T. Requarth, "Global Brain," March 3, 2016, https://www.simonsfoundation .org/2016/03/03/how-do-different-brain-regions-interact-to-enhance-func tion/.
18. J. Clack, "Distinguishing Between 'Macro' and 'Micro' Possibility Thinking: Seen and Unseen Creativity," *Thinking Skills and Creativity* 26 (2017): 60–70.
19. A. Starr, M. E. Libertus, and E. M. Brannon, "Number Sense in Infancy Predicts Mathematical Abilities in Childhood," *Proceedings of the National Academy of Sciences* 110, no. 45 (2013): 18116–20.
20. Starr et al, "Number Sense in Infancy."

21. Sakshi Gupta, "Highest Paying Data Analytics Jobs in 2024," Springboard, December 21, 2023, https://www.springboard.com/blog/data-analytics/high est-paying-analyst-jobs/.

22. J. Boaler and S. D. Levitt, "Opinion: Modern High School Math Should Be About Data Science—Not Algebra 2," *Los Angeles Times*, October 23, 2019, https://www.youcubed.org/wp-content/uploads/2019/10/LA-times-op -ed.pdf.

23. "21st Century Teaching and Learning: Data Science," youcubed.org, n.d., https://www.youcubed.org/21st-century-teaching-and-learning/.

24. "Explorations in Data Science," youcubed.org, n.d., https://hsdatascience.you cubed.org/.

25. "Application Requirements," Harvard College, n.d., https://college.harvard .edu/admissions/apply/application-requirements.

26. "Committee of Ten," Wikipedia, updated September 19, 2023, https://en.wiki pedia.org/wiki/Committee_of_Ten#:~:text=The%20National%20Education %20Association%20of,making%20recommendations%20for%20future%20 practice.

27. S. Strogatz, *Infinite Powers: How Calculus Reveals the Secrets of the Universe* (New York: Eamon Dolan Books, 2019).

28. National Center for Education Statistics, "High School Mathematics and Science Course Completion," *Condition of Education* (US Department of Education: Institute of Education Sciences, 2022), https://nces.ed.gov/programs/coe /indicator/sod/high-school-courses.

29. M. L. Hayes, *2018 NSSME+: Status of High School Mathematics* (Chapel Hill, NC: Horizon Research, 2019), http://horizon-research.com/NSSME/wp -content/uploads/2019/05/2018-NSSME-Status-of-High-School-Math.pdf.

30. D. Bressoud, ed., *The Role of Calculus in the Transition from High School to College Mathematics* (Mathematical Association of America & National Council of Teachers of Mathematics, 2017).

31. See 2023 Mathematics Framework, California Department of Education.

32. J. Ewing, "Should I Take Calculus in High School?" *Forbes*, February 15, 2020, https://www.forbes.com/sites/johnewing/2020/02/15/should-i-take-calculus -in-high-school/?sh=69ae46867625.

33. See "Explorations in Data Science," youcubed.org.

34. J. Boaler, K. Conte, K. Cor, J. Dieckmann, T. LaMar, J. Ramirez, and M. Selbach-Allen, "Studying the Opportunities Provided by an Applied High School Mathematics Course: Explorations in Data Science," *Journal of Statistics and Data Science Education* (forthcoming).

35. Jessica Furr (Waggener), "A Brief History of Mathematics Education in America," University of Georgia, spring 1996, http://jwilson.coe.uga.edu/EMAT 7050/HistoryWeggener.html.

36. R. Swartzentruber, "Data-Wisdom as a Framework for Building Data Literacy" (master's thesis, University of Tennessee, 2023), https://trace.tennessee .edu/utk_gradthes/9229.

37. For the color version of the data visuals, visit www.youcubed.org/resource/data -talks/.

38. InStat Sport, "InStat Football Webinar: Use of API with Michael Poma," YouTube.com, June 3, 2020, https://www.youtube.com/watch?app=desktop &v=qemgfwwbbPM.

39. T. Chartier, *Get in the Game: An Interactive Introduction to Sports Analytics* (Chicago: Univ. Press, 2022).

40. Dear Data, http://www.dear-data.com/.

41. Dear Data.

42. See "Explorations in Data Science," youcubed.org.

43. T. LaMar, "Data Science as a Gateway to Belonging in STEM and Other Quantitative Fields" (PhD diss., Stanford University, 2023).

44. "Spurious Correlations," tylervigen.com, n.d., https://www.tylervigen.com /spurious-correlations.

45. "Facebook–Cambridge Analytica Data Scandal," Wikipedia, n.d., https:// en.wikipedia.org/wiki/Facebook%E2%80%93Cambridge_Analytica_data _scandal.

Chapter 5: Mathematics as a Visual Experience

1. Boaler, "Prove It to Me!"

2. "Painted Cube," youcubed.org, n.d., https://www.youcubed.org/tasks/painted -cube/.

3. Menon, "Brain Networks for Mental Arithmetic."

4. Ericsson and Pool, *Peak*.

5. Ericsson and Pool, *Peak*.

6. Ericsson and Pool, *Peak*.

7. J. Hawkins, *A Thousand Brains: A New Theory of Intelligence* (New York: Basic Books, 2021).

8. L. Bofferding, "Negative Integer Understanding: Characterizing First Graders' Mental Models," *Journal for Research in Mathematics Education* 45, no. 2 (2014): 194–245; J. M. Tsang, K. P. Blair, L. Bofferding, and D. L. Schwartz, "Learning to 'See' Less Than Nothing: Putting Perceptual Skills to Work for Learning Numerical Structure," *Cognition and Instruction* 33, no. 2, (2015): 154–97; S. L. Macrine and J. M. Fugate, eds., *Movement Matters: How Embodied Cognition Informs Teaching and Learning* (Chicago: MIT Press, 2022).

9. Amalric and Dehaene, "Origins of the Brain Networks"; R. A. Cortes, E. G. Peterson, D. J. Kraemer, R. A. Kolvoord, D. H. Uttal, N. Dinh, . . . , and A. E. Green, "Transfer from Spatial Education to Verbal Reasoning and Prediction of Transfer from Learning-Related Neural Change," *Science Advances* 8, no. 31 (2022): eabo3555.

10. Menon, "Brain Networks for Mental Arithmetic."

11. C. Kalb, "What Makes a Genius?" *National Geographic*, May 2017.

12. J. Park and E. M. Brannon, "Training the Approximate Number System Improves Math Proficiency," *Psychological Science* 24, no. 10 (2013): 2013–19; Cortes et al., "Transfer from Spatial Education."

13. Bofferding, "Negative Integer Understanding"; Tsang et al., "Learning to 'See'"; Macrine and Fugate, eds., *Movement Matters*.

14. Bruce McCandliss, Stanford Graduate School of Education, n.d., https://ed.stanford.edu/faculty/brucemc.
15. M. Guillaume, E. Roy, A. Van Rinsveld, G. S. Starkey, Project iLead Consortium, M. R. Uncapher, and B. D. McCandliss, "Groupitizing Reflects Conceptual Developments in Math Cognition and Inequities in Math Achievement from Childhood through Adolescence," *Child Development* 94, no. 2 (2023): 335–47.
16. I. Benson, N. Marriott, and B. D. McCandliss, "Equational Reasoning: A Systematic Review of the Cuisenaire-Gattegno Approach," *Frontiers in Education* (2022): 507.
17. M. Penner-Wilger, L. Fast, J.-A. LeFevre, B. L. Smith-Chant, S.-L. Skwarchuk, D. Kamawar, and J. Bisanz, "Subitizing, Finger Gnosis, and the Representation of Number," *Proceedings of the 31st Annual Cognitive Science Society* 31 (2009): 520–25.
18. J. Boaler and L. Chen, "Why Kids Should Use Their Fingers in Math Class," *Atlantic*, April 13, 2016, https://www.theatlantic.com/education/archive/2016/04/why-kids-should-use-their-fingers-in-math-class/478053/.
19. R. S. Siegler and G. B. Ramani, "Playing Linear Numerical Board Games Promotes Low-Income Children's Numerical Development," *Developmental Science* 11 (2008): 655–61.
20. Yeager et al., "Teacher Mindsets Help Explain."
21. Anderson et al., "Achieving Elusive Teacher Change," 98; Bui et al., "Systematic Review."
22. I have written a series of K–8 mathematics books with coauthors Jen Munson and Cathy Williams; see "K–8 Curriculum," youcubed.org.
23. Gray and Tall, "Duality, Ambiguity, and Flexibility"; H. Chang, L. Chen, Y. Zhang, Y. Xie, C. de Los Angeles, E. Adair, . . . , and V. Menon, "Foundational Number Sense Training Gains Are Predicted by Hippocampal–Parietal Circuits," *Journal of Neuroscience* 42, no. 19 (2022): 4000–15.
24. L. Ma, *Knowing and Teaching Elementary Mathematics: Teachers' Understanding of Fundamental Mathematics in China and the United States*, Studies in Mathematical Thinking and Learning Series (Oxfordshire, UK: Routledge, 2010).
25. "Online Student Course," youcubed.org, n.d., https://www.youcubed.org/online-student-course/.
26. "WIM Videos," youcubed.org, n.d., https://www.youcubed.org/resource/wim-videos/.
27. J. Boaler and C. Humphreys, *Connecting Mathematical Ideas: Middle School Video Cases to Support Teaching and Learning* (Portsmouth, NH: Heinemann, 2005).
28. "Videos," youcubed.org, n.d., https://www.youcubed.org/resource/videos/.
29. M. Cordero, "It's (Not) Ours to Reason Why: A Comparative Analysis of Algorithms for the Division of Fractions" (honor's thesis, Stanford University, 2017), 1.
30. Duane Habecker, "Dividing Fractions with Common Denominators," YouTube.com, May 17, 2019, https://www.youtube.com/watch?v=uixRVcArQDQ; Cordero, "It's (Not) Ours to Reason Why," 1.
31. Cordero, "It's (Not) Ours to Reason Why," 1.

32. D. D. Pesek and D. Kirshner, "Interference of Instrumental Instruction in Subsequent Relational Learning," *Journal for Research in Mathematics Education* 31, no. 5 (2000): 524–40.

33. T. P. Carpenter and M. K. Corbitt, eds., "Results from the Second Mathematics Assessment of the National Assessment of Educational Progress," Fraction Bars, n.d., https://fractionbars.com/Research_Tch_Fracs/Results2nd.html.

34. "Online Student Course," youcubed.org.

35. Merzenich, *Soft-Wired*, 2; Doidge, *Brain That Changes Itself.*

36. Dweck and Yeager, "Mindsets."

37. Boaler, "Prove It to Me!"

38. Fawn Nguyen gives many examples of algebraic patterns on her lovely site, "Visual Patterns," https://www.visualpatterns.org/.

39. "Mathematical Mindset Algebra," youcubed.org, n.d., https://www.youcubed.org/algebra/.

40. A. Proehl, "For Bay Area Designer Diarra Bousso, Math + Art = Happiness," KQED, June 1, 2023, https://www.kqed.org/arts/13929878/for-bay-area-designer-diarra-bousso-math-art-happiness.

41. E. Farra, "A Senegal-Raised, Silicon Valley-Based Designer Shares Her Vision for a More Sustainable and Inclusive Future," *Vogue*, June 5, 2020, https://www.vogue.com/article/diarra-bousso-diarrablu-sustainable-made-in-senegal-collection; H. Jennings, "Meet Diarra Bousso: One of Senegal's Most Promising Designers," CNN, April 19, 2021, https://www.cnn.com/style/article/diarrablu-diarra-bousso-senegal/index.html.

42. "Diarra Bousso, MA '18 Stanford Teacher Education Program: Fusing Fashion and Math," Stanford Graduate School of Education, December 12, 2022, https://ed.stanford.edu/about/community/diarra-bousso.

43. Proehl, "For Bay Area Designer Diarra Bousso."

44. "What Is Notice and Wonder?" National Council of Teachers of Mathematics, n.d., https://www.nctm.org/noticeandwonder/.

Chapter 6: The Beauty of Mathematical Concepts and Connections

1. Gray and Tall, "Duality, Ambiguity, and Flexibility."

2. Gray and Tall, "Duality, Ambiguity, and Flexibility."

3. W. P. Thurston, "Mathematical Education" (2005): 5, arXiv.org/abs/math/0503081.

4. Paper forthcoming; check youcubed.org for updates.

5. California Digital Learning Integration and Standards Guidance, https://www.cadlsg.com/.

6. J. D. Bransford, A. L. Brown, and R. R. Cocking, *How People Learn*, vol. 11 (Washington, DC: National Academy Press, 2000), 20.

7. See 2023 Mathematics Framework, California Department of Education.

8. "K–8 Curriculum," youcubed.org.

9. Hawkins, *A Thousand Brains.*

10. "Big Picture Thinking: Definition, Strategies and Careers," Indeed, updated June 24, 2022, https://www.indeed.com/career-advice/career-development/big-picture-thinking-strategies.

11. L. Fries, J. Y. Son, K. B. Givvin, and J. W. Stigler, "Practicing Connections: A

Framework to Guide Instructional Design for Developing Understanding in Complex Domains," *Educational Psychology Review* 33, no. 2 (2021): 739–62.

12. "K–8 Curriculum," youcubed.org.

13. "Sketchnoting in the Classroom," Verbal to Visual, n.d., https://verbaltovisual.com/sketchnoting-in-the-classroom/.

14. A. Fernández-Fontecha, K. L. O'Halloran, S. Tan, and P. Wignell, "A Multimodal Approach to Visual Thinking: The Scientific Sketchnote," *Visual Communication* 18, no. 1 (2019): 5–29, at 7, https://journals.sagepub.com/doi/pdf/10.1177/1470357218759808.

15. Educators who are sharing the practice of sketchnoting and providing helpful resources for everyone include Laura Wheeler and the Verbal to Visual YouTube channel; see "Tap Into the Power of Your Visual Brain," Verbal to Visual, n.d., https://verbaltovisual.com/an-introduction-to-visual-note-taking.

16. P. A. Mueller and D. M. Oppenheimer, "The Pen Is Mightier Than the Keyboard: Advantages of Longhand over Laptop Note Taking," *Psychological Science* 25, no. 6 (2014): 1159–68.

17. A. H. Ziadat, "Sketchnote and Working Memory to Improve Mathematical Word Problem Solving Among Children with Dyscalculia," *International Journal of Instruction* 15, no. 1 (2022): 509–26; K. Fernandez and J. He, "Designing Sketch and Learn: Creating a Playful Sketching Experience That Helps Learners Build a Practice Toward Visual Notetaking (aka Sketchnotes)," Stanford Libraries Digital Stack, 2018, https://stacks.stanford.edu/file/druid:jx835yk3980/fernandez_he_sketch_and_learn.pdf.

18. M. Rohde, "Heidee Vincent Creates Sketchnotes to Help Her University Students Learn and Understand Math," *Sketchnote Army* (blog), December 7, 2020, https://sketchnotearmy.com/blog/2020/12/7/heidee-vincent-math-sketchnotes.

19. "Icons, Illustrations, Photos, Music, and Design Tools," Icons8, n.d., https://icons8.com/.

20. "How to Learn Math for Teachers," Stanford Online.

21. Anderson et al., "Achieving Elusive Teacher Change," 98.

22. Anderson et al., "Achieving Elusive Teacher Change," 98.

23. Boaler and Humphreys, *Connecting Mathematical Ideas*; C. Humphreys and R. Parker, *Making Number Talks Matter* (Grandview Heights, OH: Stenhouse Publishers, 2015); R. Parker and C. Humphreys, *Digging Deeper: Making Number Talks Matter Even More* (Grandview Heights, OH: Stenhouse Publishers, 2018).

24. M. T. Battista, "Fifth Graders' Enumeration of Cubes in 3D Arrays: Conceptual Progress in an Inquiry-Based Classroom," *Journal for Research in Mathematics Education* 30, no. 4 (1999): 417–48.

25. J. F. Shumway, *Number Sense Routines: Building Mathematical Understanding Every Day in Grades 3–5* (Grandview Heights, OH: Stenhouse Publishers, 2018).

26. Pesek and Kirshner, "Interference of Instrumental Instruction."

27. Pesek and Kirshner, "Interference of Instrumental Instruction."

28. C. Kieran, "A Comparison Between Novice and More-Expert Algebra Students on Tasks Dealing with the Equivalence of Equations," in *Proceedings*

of the Sixth Annual Meeting of the North American Chapter of the International Group for the Psychology of Mathematics Education, ed. J. M. Moser, 83–91 (Madison: Univ. of Wisconsin, 1984).

29. D. Wearne and J. Hiebert, "A Cognitive Approach to Meaningful Mathematics Instruction: Testing a Local Theory Using Decimal Numbers," *Journal for Research in Mathematics Education* 19 (1988): 371–84.

30. N. K. Mack, "Learning Fractions with Understanding: Building on Informal Knowledge," *Journal for Research in Mathematics Education* 21 (1990): 16–32.

31. Pesek and Kirshner, "Interference of Instrumental Instruction," 526.

32. Pesek and Kirshner, "Interference of Instrumental Instruction."

33. M. Cordero, M. Leshin, M. Selbach-Allen, and T. LaMar, "Exploring Calculus," youcubed.org, n.d., https://www.youcubed.org/exploring-calculus/.

34. "What Can Math Reveal About Our World and Ourselves?" Steven Strogatz, n.d., https://www.stevenstrogatz.com/.

35. Strogatz, *Infinite Powers*.

36. Strogatz, *Infinite Powers*, xiv.

37. Strogatz, *Infinite Powers*, xv.

38. Strogatz, *Infinite Powers*.

39. "The Volume of a Lemon," youcubed.org, n.d., https://www.youcubed.org/resources/the-volume-of-a-lemon/.

40. J. Boaler, K. Brown, T. LaMar, M. Leshin, and M. Selbach-Allen, "Infusing Mindset Through Mathematical Problem Solving and Collaboration: Studying the Impact of a Short College Intervention," *Education Sciences* 12 (2022): 694, https://doi.org/10.3390/educsci12100694.

41. Cordero et al., "Exploring Calculus."

42. Boaler et al., "Infusing Mindset," 694.

43. "Our People," Nk'mip Desert Cultural Centre, n.d., https://nkmipdesert.com/our-people/.

44. Clack, "Distinguishing Between 'Macro' and 'Micro' Possibility Thinking."

45. Z. Hammond, *Culturally Responsive Teaching and the Brain: Promoting Authentic Engagement and Rigor Among Culturally and Linguistically Diverse Students* (Thousand Oaks, CA: Corwin Press, 2014).

46. "Indigenous Mathematical Art," youcubed.org, n.d., https://www.youcubed.org/resource/indigenous-maths-art/.

47. For examples of rich tasks, see our K–8 book series, which sets out big ideas for each grade, and in which many authors in mathematics education share beautiful, conceptual tasks: "K–8 Curriculum," youcubed.org.

Chapter 7: Diversity in Practice and Feedback

1. "K. Anders Ericsson," Wikipedia, updated December 26, 2022, https://en.wikipedia.org/wiki/K._Anders_Ericsson.

2. Ericsson and Pool, *Peak*.

3. Ericsson and Pool, *Peak*.

4. Hecht et al., "Shifting the Mindset Culture."

5. J. Boaler, "Open and Closed Mathematics Approaches: Student Experiences and Understandings," *Journal for Research in Mathematics Education* 29, no. 1 (1998): 41–62; J. Boaler, *Experiencing School Mathematics: Traditional and Re-*

form Approaches to Teaching and Their Impact on Student Learning (Mahwah, NJ: Lawrence Erlbaum Associates, 2002).

6. Boaler, *Experiencing School Mathematics*.

7. Macrine and Fugate, eds., *Movement Matters*; L. Shapiro and S. A. Stolz, "Embodied Cognition and Its Significance for Education," *Theory and Research in Education* 17, no. 1 (2019): 19–39; D. Abrahamson and A. Bakker, "Making Sense of Movement in Embodied Design for Mathematics Learning," *Cognitive Research: Principles and Implications* 1, no. 1 (2016): 1–13; D. Abrahamson, "Embodied Design: Constructing Means for Constructing Meaning," *Educational Studies in Mathematics* 70 (2009): 27–47.

8. K. P. Blair, M. Rosenberg-Lee, J. M. Tsang, D. L. Schwartz, and V. Menon, "Beyond Natural Numbers: Negative Number Representation in Parietal Cortex," *Frontiers in Human Neuroscience* 6 (2012): 7.

9. D. L. Schwartz, J. M. Tsang, and K. P. Blair, *The ABCs of How We Learn: 26 Scientifically Proven Approaches, How They Work, and When to Use Them* (New York: W. W. Norton, 2016); Schwartz and Bransford, "Time for Telling"; S. Levine, "Contrasting Cases: A Simple Strategy for Deep Understanding," Cult of Pedagogy, March 20, 2022, https://www.cultofpedagogy.com/contrasting-cases/.

10. Schwartz et al., *ABCs of How We Learn*; Schwartz and Bransford, "Time for Telling"; Levine, "Contrasting Cases."

11. Schwartz et al., *ABCs of How We Learn*; Schwartz and Bransford, "Time for Telling"; Levine, "Contrasting Cases."

12. Schwartz et al., *ABCs of How We Learn*; Schwartz and Bransford, "Time for Telling"; Levine, "Contrasting Cases."

13. H. Luo, T. Yang, J. Xue, and M. Zuo, "Impact of Student Agency on Learning Performance and Learning Experience in a Flipped Classroom," *British Journal of Educational Technology* 50, no. 2 (2019): 819–31; P. Wiliams, "Student Agency for Powerful Learning," *Knowledge Quest* 45, no. 4 (2017): 8–15; J. Boaler and T. Sengupta-Irving, "The Many Colors of Algebra: The Impact of Equity Focused Teaching upon Student Learning and Engagement," *Journal of Mathematical Behavior* 41 (2016): 179–90; J. Arnold and D. J. Clarke, "What Is 'Agency'? Perspectives in Science Education Research," *International Journal of Science Education* 36, no. 5 (2014): 735–54; J. Boaler and J. G. Greeno, "Identity, Agency, and Knowing," *Multiple Perspectives on Mathematics Teaching and Learning* 1 (2000): 171.

14. J. Boaler and S. K. Selling, "Psychological Imprisonment or Intellectual Freedom?: A Longitudinal Study of Contrasting School Mathematics Approaches and Their Impact on Adults' Lives," *Journal for Research in Mathematics Education* 48, no. 1 (2017): 78–105.

15. Boaler, *Experiencing School Mathematics*; Boaler, "Open and Closed Mathematics."

16. Boaler, *Experiencing School Mathematics*.

17. Boaler and Selling, "Psychological Imprisonment or Intellectual Freedom?"

18. G. Hatano and Y. Oura, "Commentary: Reconceptualizing School Learning Using Insight from Expertise Research," *Educational Researcher* 32, no. 8 (2003): 26–29.

19. Boaler and Selling, "Psychological Imprisonment or Intellectual Freedom?"

20. M. Suri, "Declines in Math Readiness Underscore the Urgency of Math Awareness," The 74, April 5, 2023, https://www.the74million.org/article/declines-in-math-readiness-underscore-the-urgency-of-math-awareness/.

21. M. D. Felton, C. O. Anhalt, and R. Cortez, "Going with the Flow: Challenging Students to Make Assumptions," *Mathematics Teaching in the Middle School* 20, no. 6 (2015): 342–49.

22. "Excerpt of Jo from 'The Importance of Struggle,'" youcubed.org; "The Importance of Struggle," youcubed.org.

23. Boaler, *Mathematical Mindsets*; Hecht et al., "Shifting the Mindset Culture."

24. "Wolfram Mathematica," Wolfram, n.d., https://www.wolfram.com/mathematica/.

25. "WolframAlpha," Wolfram, n.d., https://www.wolframalpha.com/.

26. C. Wolfram, "Teaching Kids Real Math with Computers," TED, July 2010, www.ted.com/talks/conrad_wolfram_teaching_kids_real_math_with_computers?language=en.

27. "Let's Fix Maths Education," computerbasedmath.org, n.d., https://www.computerbasedmath.org/.

28. E. L. Bjork and R. A. Bjork, "Making Things Hard on Yourself, But in a Good Way: Creating Desirable Difficulties to Enhance Learning," *Psychology and the Real World: Essays Illustrating Fundamental Contributions to Society* 2 (2011): 59–68.

29. "Carl Wieman," Wikipedia, updated October 18, 2023, https://en.wikipedia.org/wiki/Carl_Wieman.

30. C. Wieman, "Why Not Try a Scientific Approach to Science Education?" *Change: The Magazine of Higher Learning* 39, no. 5 (2007): 9–15.

31. "Carl Wieman," Stanford Profiles, n.d., https://profiles.stanford.edu/carl-wieman.

32. L. Deslauriers, E. Schelew, and C. Wieman, "Improved Learning in a Large-Enrollment Physics Class," *Science* 332, no. 6031 (2011): 862–64.

Chapter 8: A New Mathematical Future

1. Bryan et al., "Shifting the Mindset Culture."

2. Lawyers' Committee for Civil Rights of the San Francisco Bay Area, *Held Back: Addressing Misplacement of 9th Grade Students in Bay Area School Math Classes*, January 2013, https://lccrsf.org/wp-content/uploads/HELD-BACK-9th-Grade-Math-Misplacement.pdf.

3. 2023 Mathematics Framework, California Department of Education.

4. Boaler and Staples, "Creating Mathematical Futures"; Boaler, "Open and Closed Mathematics."

5. Drew, "Why Science Majors Change Their Minds."

6. S. Clivaz and T. Miyakawa, "The Effects of Culture on Mathematics Lessons: An International Comparative Study of a Collaboratively Designed Lesson," *Educational Studies in Mathematics* 105, no. 1 (2020): 53–70.

7. Deslauriers et al., "Measuring Actual Learning"; Kapur, "Productive Failure in Learning Math"; Schwartz et al., "Practicing Versus Inventing," 759; Schwartz and Bransford, "Time for Telling."

8. Deslauriers et al., "Improved Learning."

9. Boaler et al., "Transformative Impact," 512.

10. Alexei is a senior lecturer at the University of Essex, a position that translates to a "professor" in the US.

11. I. Daly, J. Bourgaize, and A. Vernitski, "Mathematical Mindsets Increase Student Motivation: Evidence from the EEG," *Trends in Neuroscience and Education* 15 (2019): 18–28.

12. Daly et al., "Mathematical Mindsets Increase Student Motivation."

13. A. L. Campbell, M. Mokhithi, J. P. Shock, "Exploring Mathematical Mindset in Question Design: Boaler's Taxonomy Applied to University Mathematics," in *REES AAEE 2021 Conference: Engineering Education Research Capability Development*, 980–88 (Perth, WA: Engineers Australia, 2021).

14. "Sol Garfunkel," Wikipedia, n.d., https://en.wikipedia.org/wiki/Sol_Garfunkel.

15. See Consortium for Mathematics and Its Applications, https://www.comap .com/.

16. See "The Mathematical Contest in Modeling (MCM) / The Interdisciplinary Contest in Modeling (ICM)," Consortium for Mathematics and Its Applications, https://www.comap.com/contests/mcm-icm.

17. See International Mathematical Olympiad, https://www.imo-official.org/.

18. Boaler, "Paying the Price for 'Sugar and Spice.'"

19. A. K. Whitney, "Math for Girls, Math for Boys," *Atlantic*, April 18, 2016, https://www.theatlantic.com/education/archive/2016/04/girls-math-interna tional-competition/478533/.

20. "William Lowell Putnam Mathematical Competition," Wikipedia, updated October 1, 2023, https://en.wikipedia.org/wiki/William_Lowell_Putnam _Mathematical_Competition#:~:text=It%20is%20widely%20considered%20 to,by%20students%20specializing%20in%20mathematics.

21. "William Lowell Putnam Mathematical Competition," Mathematical Association of America, n.d., https://www.maa.org/sites/default/files/pdf/Putnam /Competition_Archive/List%20of%20Previous%20Putnam%20Winners.pdf.

22. J. Boaler, M. Cordero, and J. Dieckmann, "Pursuing Gender Equity in Mathematics Competitions: A Case of Mathematical Freedom," *MAA Focus* (February/March 2019), http://digitaleditions.walsworthprintgroup.com/pub lication/?m=7656&l=1&i=566588&p=18&ver=html5.

23. See Consortium for Mathematics and Its Applications, https://www.comap .com/contests/mcm-icm.

24. Boaler et al., "Pursuing Gender Equity."

25. Boaler et al., "Pursuing Gender Equity."

26. T. Grandin, *Visual Thinking: The Hidden Gifts of People Who Think in Pictures, Patterns, and Abstractions* (New York: Penguin, 2022).

27. S. M. Iversen and C. J. Larson, "Simple Thinking Using Complex Math Vs. Complex Thinking Using Simple Math—A Study Using Model Eliciting Activities to Compare Students' Abilities in Standardized Tests to Their Modelling Abilities," *ZDM* 38 (2006): 281–92; Boaler et al., "Studying the Opportunities."

28. Anderson et al., "Achieving Elusive Teacher Change," 98.

29. "Tai-Danae Bradley," youcubed.org, n.d., https://www.youcubed.org/resources/tai-danae-bradley/.
30. "Research Articles," youcubed, n.d., https://www.youcubed.org/evidence/research-articles/.
31. Boaler, "Promoting 'Relational Equity.'"
32. Boaler and Greeno, "Identity, Agency, and Knowing," 171.
33. Boaler and Greeno, "Identity, Agency, and Knowing."
34. Boaler and Sengupta-Irving, "The Many Colors of Algebra."
35. Cheng, "What If Nobody Is Bad at Maths?"; E. Cheng, *Is Math Real? How Simple Questions Lead Us to Mathematics' Deepest Truths* (New York, Basic Books, 2023).
36. "The Interactive Mathematics Program (IMP)," Activate Learning, n.d., https://activatelearning.com/interactive-mathematics-program-imp/.
37. Grandin, *Visual Thinking*.
38. Boaler, "Crossing the Line."
39. Boaler and Staples, "Creating Mathematical Futures."
40. Boaler, "Open and Closed Mathematics."
41. Boaler, "Crossing the Line."
42. "Tucker Carlson," Wikipedia, updated October 20, 2023, https://en.wikipedia.org/wiki/Tucker_Carlson.
43. A. Oksanen, M. Celuch, R. Latikka, R. Oksa, and N. Savela, "Hate and Harassment in Academia: The Rising Concern of the Online Environment," *Higher Education* 84 (2022): 541–67, https://doi.org/10.1007/s10734-021-00787-4.
44. M. V. Valero, "Death Threats, Trolling, and Sexist Abuse: Climate Scientists Report Online Attacks," *Nature*, April 6, 2023, https://www.nature.com/articles/d41586-023-01018-9.
45. J. V. Chamary, "Wikipedia's 100 Most Controversial People," *Forbes*, January 25, 2016, https://www.forbes.com/sites/jvchamary/2016/01/25/wikipedia-people/?sh=5522df036ffb.
46. "Jo Boaler," Wikipedia, updated August 14, 2023, https://en.wikipedia.org/wiki/Jo_Boaler.
47. L. S. Shulman, "PCK: Its Genesis and Exodus," in *Re-Examining Pedagogical Content Knowledge in Science Education*, 13–23 (Oxfordshire, UK: Routledge, 2015).
48. D. L. Ball and D. K. Cohen, "Developing Practice, Developing Practitioners: Toward a Practice-Based Theory of Professional Education," Teaching as the Learning Profession: Handbook of Policy and Practice 1 (1999): 3–22.
49. J. Boaler, "Educators, You're the Real Experts. Here's How to Defend Your Profession," Education Week, November 3, 2022, https://www.edweek.org/teaching-learning/opinion-educators-youre-the-real-experts-heres-how-to-defend-your-profession/2022/11.
50. L. A. Santos, J. G. Voelkel, R. Willer, and J. Zaki, "Belief in the Utility of Cross-Partisan Empathy Reduces Partisan Animosity and Facilitates Political Persuasion," *Psychological Science* 33, no. 9 (2022), https://doi.org/10.1177/09567976221098594.

51. I. Manji, *Don't Label Me: An Incredible Conversation for Divided Times* (New York: St. Martin's Press, 2019).
52. J. Little, *The Warrior Within: The Philosophies of Bruce Lee* (New York: Chartwell Books, 2016).
53. Boaler, *Limitless Mind*.
54. G. Lukianoff and J. Haidt, *The Coddling of the American Mind: How Good Intentions and Bad Ideas Are Setting Up a Generation for Failure* (New York: Penguin, 2019).
55. Boaler, "Crossing the Line."
56. "An Example of a Growth Mindset K–8 School," youcubed.org, n.d., https://www.youcubed.org/resources/an-example-of-a-growth-mindset-k-8-school/.
57. C. Trungpa, *Shambhala: The Sacred Path of the Warrior* (Boulder, CO: Shambhala Publications, 2009).
58. "Bruce Lee," Wikipedia, updated October 17, 2023, https://en.wikipedia.org/wiki/Bruce_Lee.
59. Little, *Warrior Within*, xxii.
60. K. Danaos, *Nei Kung: The Secret Teachings of the Warrior Sages* (New York: Simon and Schuster, 2002).
61. Little, *Warrior Within*, xxii.
62. Trungpa, *Shambhala*, 262.
63. K. P. Ellison, *Untangled: Walking the Eightfold Path to Clarity, Courage, and Compassion* (New York: Balance Books, 2022), 62.